SO-AIT-892

Transform Analysis
and Filters

Transform Analysis and Filters

LEONARD J. GEIS

Professor of Electronics Engineering Technology
DeVry Institute of Technology
Lombard, Illinois

Prentice Hall, *Englewood Cliffs, New Jersey 07632*

Library of Congress Cataloging-in-Publication Data

Geis, Leonard J.
  Transform analysis and filters.

  Bibliography: p.
  Includes index.
  1. Electric filters, Passive. 2. Electric filters,
Active. 3. Electric filters, Digital. 4. Switched
capacitor circuits. 5. Laplace transformation.
I. Title.
TK7872.F5G425 1989   621.3815′324   88-6030
ISBN 0-13928912−7

Editorial/production supervision and
interior design: Lillian Glennon
Cover design: Diane Saxe
Manufacturing buyer: Bob Anderson

© 1989 by Prentice-Hall, Inc.
A Division of Simon & Schuster
Englewood Cliffs, New Jersey 07632

Printed in the United States of America
10 9 8 7 6 5 4 3 2 1

ISBN 0-13-928912-7

PRENTICE-HALL INTERNATIONAL (UK) LIMITED, *London*
PRENTICE-HALL OF AUSTRALIA PTY. LIMITED, *Sydney*
PRENTICE-HALL CANADA INC., *Toronto*
PRENTICE-HALL HISPANOAMERICANA, S.A., *Mexico*
PRENTICE-HALL OF INDIA PRIVATE LIMITED, *New Delhi*
PRENTICE-HALL OF JAPAN, INC., *Tokyo*
SIMON & SCHUSTER ASIA PTE. LTD., *Singapore*
EDITORA PRENTICE-HALL DO BRASIL, LTDA., *Rio de Janeiro*

contents

preface

This book is meant to be used as a textbook in a junior-level course in a college engineering technology curriculum, to give students an understanding of filters, their use, analysis and synthesis. Because a filter's response varies with frequency, it will become convenient to express the results in terms of the frequency variable. When working with frequency as a variable, Laplace transforms become a very useful tool. Because Laplace transforms will handle other mathematical functions besides sinusoidal functions, a more general analysis of circuits becomes possible. For this reason, the first chapter deals with step, ramp, and impulse functions and their Laplace transforms, as well as sinusoidal time functions and their Laplace transforms. The second chapter applies these tools to linear, passive, time-invariant circuits.

The second part of this book covers analysis of Butterworth, Chebyshev, elliptic, and Bessel filters, both passive and active. Synthesis of these filters is performed by the use of nomographs, tables, standard schematics, and frequency and impedance scaling methods. Conversion techniques are given for obtaining high-pass, bandpass, and band-reject filters from a prototype low-pass filter.

The third part of this book is an introduction to two related topics. A chapter on switched capacitors introduces the student to a low-cost method of realizing accurate and inexpensive audio filters. Finally, z-transforms are introduced for use in analyzing digital or discrete filters, which are becoming increasingly important as technology is converting to a digital world. It is assumed that the student already has a background in algebra, trigonometry, differential and integral calculus and dc and ac circuit analysis.

The incentive for writing this book came from the need to combine into one book the information presently being taught in a 4-hour course at the DeVry schools. Because it is taught in the electronics engineering technology curriculum, it was desired to keep the mathematics to a minimum; but where it is necessary, the mathematical steps are carried through in detail. About half of the course is spent on Chapters 1 and 2. Chapters 3 through 5 move somewhat faster, since much of the material is almost of a "cookbook" technique of using nomographs, tables, and "plug-in" formulas. Chapters 6 and 7 are usually just briefly discussed in class. If the students were already exposed to Laplace transforms in a previous math course, much time could be saved by starting with Chapter 2, allowing more depth of coverage for the later chapters. Another strategy would be to split the course into two 3-hour courses: the first, called "Transform Methods," would cover the first two chapters and part of the last chapter; the second, called "Filter Theory," would cover the rest of the material.

Each chapter begins with a list of learning objectives, followed by an introduction which shows how that chapter's content fits into the overall scheme of the book. Each chapter ends with a summary of the important points discussed, followed by a set of problems arranged to correspond to the chapter sections.

The back of the book contains several sections of value to the student. Nothing is more frustrating than spending a great deal of time solving a problem and not knowing if the answer is right. So one section at the back contains answers to chapter problems. Another section is a glossary of new terms. These are terms that appear in boldfaced type the first time they are encountered in the text. The appendices contain useful formulas and several BASIC computer programs. They enable the student to factor polynomials, do partial fraction expansion, and do a frequency analysis and Bode plot of the voltage transfer ratio (common ground) of any analog circuit. Results are tabulated or plotted on a monitor or on a printer. This can be quite helpful in confirming that a filter which has been synthesized by the student actually does meet the specifications stated in the problem. A bibliography and an index are also included.

I wish to thank my wife, Reinhilde, whose patience and encouragement were vital in completing this book in a timely manner. The comments of my students and the reviewers of the manuscript were also very helpful.

Leonard J. Geis

Transform Analysis
and Filters

1

laplace transforms

OBJECTIVES

After completing this chapter, you should be able to:

- Define the step, ramp, and impulse functions.
- Resolve a graphed function into a sum or product of the functions above, or plot a time expression containing these.
- Define the Laplace transform formula and use it to find the Laplace transform of some simple time functions.
- Identify the correct transformation for a common function of time, given a table of Laplace transforms, and apply the linearity property to find the Laplace transform.
- Expand a given $F(s)$ having either distinct real poles, multiple real poles, simple complex poles, or multiple complex poles, or some combination of these into a sum of simple partial fractions and find its inverse Laplace transform.
- Express a time-shifted function in the time domain and, using tables, find the proper Laplace transform of it.

- Use tables to find the Laplace transform of a given periodic time function.
- Find the Laplace transform of the derivative of a given function if the transform of the given function is known.
- Find the Laplace transform of the integral of a given function if the transform of the given function is known.
- Determine the initial and final value of a time function if only the function's transform is known.

INTRODUCTION

In the days before calculators, to evaluate

$$\frac{52(0.087)}{23 \times 10^4(1.83)}$$

to a degree of accuracy better than could be obtained with a slide rule would require tedious multiplication and long division. To shorten this process, the technique of logarithmic transforms was devised. This reduced the problem to looking up each number's logarithm in a table, adding or subtracting these values by hand, and looking up the answer by using the table of logarithms in a reverse manner. For instance, in the problem given above, the procedure and result would look as shown in Table 1.1. At that time this was considered a great timesaver, although the tables required for seven-place accuracy would fill a book.

    In circuit analysis we also come across calculations that are quite tedious. They usually involve inductors and capacitors. Sometimes they involve just one frequency, but sometimes we must consider the response of the circuit to many frequencies. This involves the rules of complex algebra, since j is involved. In still other cases we want to know how a circuit behaves shortly after it is turned

TABLE 1.1 TABULAR LOGARITHM
PROCEDURE

n	Log n	
52	$0.71600 + 1$	Add
0.087	$0.93952 - 2$	these
23×10^4	$0.36173 + 5$	Minus
1.83	0.26245	these
1.0748×10^{-5}	$0.03134 - 5$	Result

on. The fundamental equations that describe all of these situations are integro-differential equations such as

$$V_L = L \frac{di}{dt} \quad \text{and} \quad V_C = \frac{1}{C} \int I_C \, dt \qquad (1.1)$$

If these relations are used when writing the necessary mesh or nodal equations, the calculus solution of the equation that results when the determinants are evaluated, including the boundary conditions, becomes quite tedious.

There is a technique that shortens the work quite a bit—called Laplace transforms. Of course, Laplace transforms has nothing to do with logarithms except that logarithms provide a shorthand method of handling the multiplication and division of numbers, and Laplace transforms constitute a shorthand method of handling integral or differential equations.

Because Laplace transforms will handle circuits containing sources other than steady-state sinusoidal ac, we can now expand our horizon to consider other possible sources, such as square wave, triangular wave, exponentials, transient behavior, and so on. In this section we show how these other waveforms can be described mathematically.

1.1 STEP FUNCTION

A voltage or current undergoes a step when at some point in time there is a sudden jump in value. This is usually caused by a switch turning on or off. For instance, if a circuit function (a voltage or current) behaves as shown in Figure 1.1a, this is called a **unit step**, abbreviated, $u(t)$. Its formal definition is

$$u(\text{arg}) = \begin{cases} 0 & \text{if arg} < 0 \\ 1 & \text{if arg} \geq 0 \end{cases} \qquad (1.2)$$

The functional notation, $u(\text{arg})$, just means that the value of u depends on the value of arg, the argument. The argument is merely the item inside the parentheses. Usually, the argument is t, the time, or some simple function of time. For instance, if the switch in Figure 1.1b closes at $t = 0$ and stays closed, then $v_R = u(t)$ and Figure 1.1a is the correct plot of $v_R(t)$. So, in this case, $v_R = u(t)$. Thus the letter u will be used only to indicate a step function. If $f_1(t) = u(-t)$ and $f_2(t) = -u(-t)$, then using the formal definition of $u(\text{arg})$, we see that $f_1(t)$

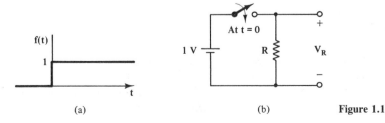

(a)   (b)   **Figure 1.1**

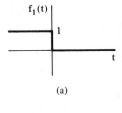

(a)

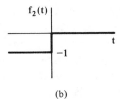

(b)   **Figure 1.2**

and $f_2(t)$ will plot as shown in Figure 1.2a and b, respectively. Students are encouraged to plot $f_1(t)$ and $f_2(t)$ themselves by plugging various values of t into the equations.

To get larger steps it is merely necessary to multiply the unit step by a constant. Also, by subtracting constants from t and using this as the argument, it is possible to create delayed steps.

Example 1.1

Plot $5u(t - 3)$ and $5u(4t - 12)$.

Solution   These are both steps 5 units high and according to the formal definition, the argument in both cases is ≥ 0 when $t \geq 3$. Therefore, both functions plot identically as shown in Figure E1.1.

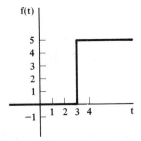

Figure E1.1

1.2 PULSE FUNCTION

If a delayed version of a step function is subtracted from the original function, a **pulse function** is created. For instance, in Figure 1.3a, $i_1(t) = 2u(t - 3)$ and in Figure 1.3b, $i_2(t) = 2u(t - 5)$. If $i_3 = i_1 - i_2$, then combining the two plots graphically will yield Figure 1.3c. This means that mathematically, $i_3(t) =$

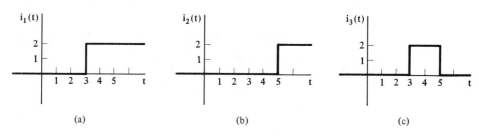

Figure 1.3

$2u(t - 3) - 2u(t - 5)$. Thus the definition of a pulse would be: a function that is turned on at some time and then turned off later. Show that another way of describing a single pulse such as $i_3(t)$ would be

$$i_3(t) = 2u(t - 3)u(-t + 5): \text{ multiplication of step functions} \qquad (1.3)$$

Combining several steps will give us the possibility of producing a great variety of stair-case functions.

Example 1.2

Describe mathematically the voltage shown in Figure E1.2.

Solution   This is a step down, two steps up, and a step down. Working from left to right, we have

$$f(t) = -2u(t + 3) + 4u(t) + 2u(t - 4) - 2u(t - 9)$$

This waveform could also be thought of as three pulses and a step.

First pulse (from $t = -3$ to $t = 0$): $-2u(t + 3) + 2u(t)$
Second pulse (from $t = 0$ to $t = 4$): $2u(t) - 2u(t - 4)$
Third pulse (from $t = 4$ to $t = 9$): $4u(t - 4) - 4u(t - 9)$
Step (from $t = 9$ to infinity): $+2u(t - 9)$

Thus $f(t)$ is the sum of these four items. It is seen that after collecting like terms, the original value for $f(t)$ is obtained.

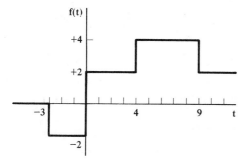

Figure E1.2

Example 1.3

Describe mathematically $v_1(t)$ for the circuit shown in Figure E1.3a and plot $v_1(t)$.

Solution   Using circuit analysis, we find that:

For $t < 4$: $v_1(t) = 1.5$ V

For $t \geq 4$: $v_1(t) = -5$ V

This means that $v_1(t)$ will plot as shown in Figure E1.3b and $v_1(t) = 1.5 - 6.5u(t - 4)$.

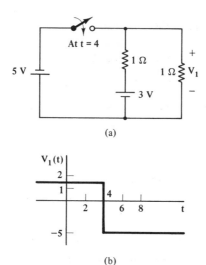

(a)

(b)   **Figure E1.3**

If a time function consists of step functions and is periodic, it is assumed that the pattern keeps repeating itself forever.   For instance, the square wave shown in Figure 1.4 starts at $t = 0$ and keeps going to $t = $ infinity.   Thus it contains an infinite number of pulses and is described as

$$f(t) = u(t) - u(t - 1) + u(t - 2) - u(t - 3) + \cdots$$

The shorthand notation for this infinite series is

$$f(t) = \sum_{n=0}^{\infty} (-1)^n u(t - n)$$

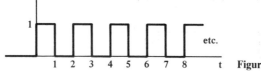

etc.

Figure 1.4

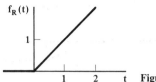

Figure 1.5

1.3 RAMP FUNCTION

Another important function is the **ramp function**. The basic ramp function is $tu(t)$, as shown in Figure 1.5. The slope is 1. Of course, other slopes are possible merely by multiplying the function above by a suitable constant. Figure 1.6a shows the function, $f_1(t) = -\frac{1}{2}tu(t)$. By delaying the step function portion, the function $-\frac{1}{2}tu(t - 1)$ of Figure 1.6b is created. This is the same as Figure 1.6a except that the value is zero for t less than 1. If we plot $-\frac{1}{2}(t - 1)u(t)$ in Figure 1.6c, we see that the function begins at $t = 0$, but it is a different function: It is the straight line going through the horizontal axis at $t = 1$. Finally, in Figure 1.6d the function $-\frac{1}{2}(t - 1)u(t - 1)$ is plotted. This amounts to a shifted version of the first function. Thus $f_4(t) = f_1(t - 1)$, which means that the t in the f_1 function is replaced by $t - 1$. Note carefully the difference in the four plots and the difference in the four mathematical expressions. To convince yourself, try plotting a few points for each expression.

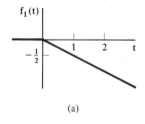

(a)

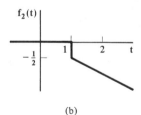

(b)

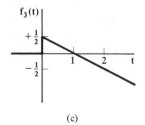

(c)

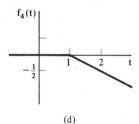

(d)

Figure 1.6

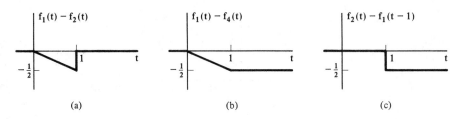

Figure 1.7

1.4 COMBINATIONS OF FUNCTIONS

Certainly, other common functions, such as $f(t) = t^2$ (a parabola) or the sine wave, $f(t) = \sin t$, could be multiplied by $u(t)$ to turn them "on." These functions could also be multiplied by constants or shifted. Finally, several of these could be added together to produce an almost endless variety of waveshapes.

By subtracting a function from itself it will disappear completely. However, if somewhat different functions are subtracted from each other, some new functions can be created. Figure 1.7 illustrates various combinations of the functions of Figure 1.6. The effect in Figure 1.7a is to have $f_1(t)$ "on" for only 1 second. We came across something similar in our discussion of pulses. The combination in Figure 1.7c reduces to a simple delayed step function. The mathematical proof is as follows:

$$f_2(t) - f_1(t - 1) = \frac{-t}{2} u(t - 1) + \frac{t - 1}{2} u(t - 1)$$

$$= -\frac{t}{2}u(t - 1) + \frac{t}{2} u(t - 1) - \frac{1}{2} u(t - 1) \qquad (1.4)$$

$$= -\frac{1}{2} u(t - 1) \qquad \text{a delayed step}$$

Example 1.4

Find an expression for Figure E1.4a.

Solution   Working from left to right, the first section is $f_1(t) = tu(t)$. At $t = 1$ second, we do not want to have any sudden jumps in value, only a change in direction, similar to Figure 1.7b, only more so. Thus we will subtract twice the shifted function from the original. This will give us the proper negative slope, but it will keep going negative after $t = 2$. Therefore, at $t = 2$ we must add another shifted function, $f_1(t - 2)$. This will just cancel the negative slope and bring the plot horizontal again. Thus

$$f_T = f_1(t) - 2f_1(t - 1) + f_1(t - 2),$$

or

$$f_T = tu(t) - 2(t - 1)u(t - 1) + (t - 2)u(t - 2)$$

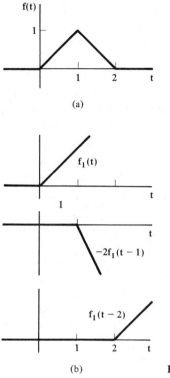

(a)

(b)   **Figure E1.4**

As a check we will plot all three components as in Figure E1.4b and add the three point by point at several places. The resulting plot should be identical with Figure E1.4a.

Example 1.5

Find an expression for Figure E1.5.

Solution   The first part looks sinusoidal or perhaps parabolic. If sinusoidal, it is either

$$-(\sin \pi t)[u(t - 1) - u(t - 2)]$$

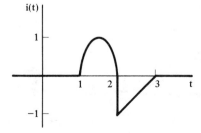

Figure E1.5

or

$$[\sin \pi(t - 1)][u(t - 1) - u(t - 2)]$$

If it is a parabola, it would be

$$-4(t - 2)(t - 1)[u(t - 1) - u(t - 2)]$$

In either case, the $u(t - 1)$ factor turns it on and the $u(t - 2)$ factor turns it off. The second section is a straight line whose equation is $t - 3$, and it is turned on at $t = 2$ and turned off at $t = 3$. Therefore, assuming a sine wave,

$$i(t) = -(\sin \pi t)[u(t - 1) - u(t - 2)] + (t - 3)[u(t - 2) - u(t - 3)]$$

Example 1.6

For Figure E1.6a find an expression for $v_c(t)$ and plot it.

Solution Even though the capacitor is connected to the voltage source for all time before $t = 1$, the source has no voltage before $t = 0$, so that for the first section,

$$v_c = \left(\sin \frac{\pi t}{2}\right)[u(t) - u(t - 1)]$$

At $t = 1$, the capacitor begins to discharge through the resistor. Since at this time the voltage is 1 V, it would seem that the voltage v_c, from circuit analysis, would be $e^{-t}u(t)$. However, this equation is good only if the discharge starts at $t = 0$. We must shift this function to the right by 1 second. Therefore, for this second section the equation becomes

$$v_c = [e^{-(t-1)}]u(t - 1)$$

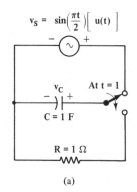

(a)

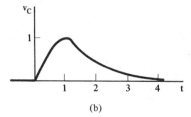

(b)   **Figure E1.6**

Therefore, the complete description is

$$v_c(t) = \left(\sin \frac{\pi t}{2} \right)[u(t) - u(t - 1)] + [e^{-(t-1)}]u(t - 1)$$

The plot is shown in Figure E1.6b.

1.5 IMPULSE FUNCTION

Pulses come in various heights and widths. In Figure 1.3 the pulse had a height of 2 and a width of 2, making the area equal to 4. If the area were equal to 1, it would be called a **unit pulse**. Figure 1.8a shows a narrow pulse that happens to be a unit pulse, since its area is $w(1/w) = 1$. If the width is made extremely small and the height proportionately larger, we still have a unit pulse, but the integral, instead of looking like Figure 1.8b, will be the unit step. When this occurs, we call it the **unit impulse**. Thus the unit impulse is the limit of the unit pulse as the pulse width approaches zero. The special symbol for a unit impulse is $\delta(t)$. Just as for the unit step, the unit impulse occurs where the argument is zero; at all other places, however, the function is zero. Since the height of the pulse is infinite, we must use a special plotting symbol. It is an arrow pointing upward and is 1 unit high. It is shown in Figure 1.9. The relationship between the step and the impulse is as follows:

$$u(t) = \int_{-\infty}^{t} \delta(t) \qquad \text{or} \qquad \delta(t) = \frac{du(t)}{dt} \qquad (1.5)$$

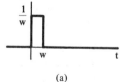

(a)

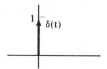

(b)   **Figure 1.8**

Figure 1.9

Example 1.7

Plot

$$f(t) = \frac{d}{dt}[-3u(t-2)]$$

Solution   The derivative of this step is an impulse.   However, it is not a unit impulse; it is negative, three times normal size, and located at $t = 2$.   It is plotted as in Figure E1.7.

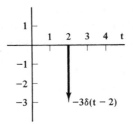

Figure E1.7

Example 1.8

In Figure E1.8a, the switch is moved to position 2 at $t = 2$.   Find an expression for i_c and plot it.

Solution   Before $t = 2$ there is no current and no voltage across the capacitor.   As the switch is closed the sinusoidal source is at its peak value of 1 V.   This means that the capacitor will instantly be charged to 1 V.   Thus the charge on the capacitor suddenly jumps from 0 to VC coulombs $= 1(2) = 2$ C.   In other words, a step of 2 C since

$$i_c = \frac{dq}{dt} \qquad i_c = 2\delta(t)$$

But since this occurs at $t = 2$, the proper notation is $i_c(t) = 2\delta(t - 2)$.   After $t = 2$, $v_c = v_s$.   Since $i_c = C\,(dv_c/dt)$, after $t = 2$,

$$i_c = 2\frac{\pi}{4}\left(\cos\frac{\pi t}{4}\right)u(t-2)$$

Therefore, the complete answer is

$$i_c = 2\delta(t-2) + \frac{\pi}{2}\left(\cos\frac{\pi t}{4}\right)u(t-2)$$

The plot is shown in Figure E1.8b.

Multiplying an impulse by a function of time will simplify to the impulse multiplied by that time function's value at the time where the impulse takes place. Thus

$$\frac{t}{2}\,\delta(t-6) = 3\delta(t-6)$$

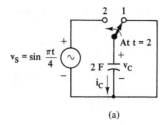

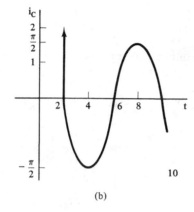

(a)

(b)   **Figure E1.8**

1.6 DEFINITION OF LAPLACE TRANSFORM

The technique of Laplace is to convert equations involving time and integrals and derivatives of time into equations involving a new variable, s, and no derivatives or integrals. The resulting equations should be easier to solve since they involve algebra but no calculus. Then by a reverse transformation, the solution could be converted back to the time domain. But what formula shall we use to convert our time function, $f(t)$, to our s function, $F(s)$? Note that we will use a lowercase letter for the time function, and the same letter, but uppercase, for our s function. Consider the following definition:

$$F(s) = \int_{0-}^{\infty} f(t)e^{-st}\, dt \qquad (1.6)$$

This is called a **one-sided Laplace transform**. If the limits of integration were from $-\infty$ to $+\infty$, it would be called a **two-sided Laplace transform**. We consider only the one-sided Laplace transform. The limit $0-$ means just slightly less than 0.

Note that the multiplying factor, e^{-st}, contains both variables s and t. This accounts for the fact that for every $f(t)$, only one $F(s)$ exists; and for every $F(s)$, only one $f(t)$ exists (for $t \geq 0$), except at possible isolated points. However, s is treated as a constant during the integration process. As a result of the integration,

t disappears, which is exactly what we want. The symbol for the process of transformation is $\mathscr{L}$. Thus

$$\mathscr{L}[f(t)] = F(s) \tag{1.7}$$

The Laplace properties are similar to the properties of integration. For instance,

$$\mathscr{L}[kf(t)] = kF(s) \tag{1.8}$$

which is proven as follows:

$$\mathscr{L}[kf(t)] = \int_0^\infty kf(t)e^{-st}\,dt = k\int_0^\infty f(t)e^{-st}\,dt = kF(s)$$

Also, the Laplace of the sum of two functions or more is the sum of the Laplace transforms of the individual functions:

$$
\begin{aligned}
\mathscr{L}[f_1(t) + f_2(t)] &= \int_0^\infty [f_1(t) + f_2(t)]e^{-st}\,dt \\
&= \int_0^\infty f_1(t)e^{-st}\,dt + \int_0^\infty f_2(t)e^{-st}\,dt \\
&= \mathscr{L}[f_1(t)] + \mathscr{L}[f_2(t)] = F_1(s) + F_2(s)
\end{aligned}
\tag{1.9}
$$

However, $\mathscr{L}[f_1(t)f_2(t)] \neq F_1(s)F_2(s)$ because $f_1(t)f_2(t)e^{-st}$ cannot be integrated or simplified unless we know the expressions for $f_1(t)$ and $f_2(t)$. These three relations, known as the **linearity property**, are a result of the rules for integration and become very important when we try to handle more complicated time functions later.

1.7 LAPLACE TRANSFORM PAIRS

Using the definition of Laplace, let us find some basic transforms.

Example 1.9

Find $\mathscr{L}[1]$.

Solution

$$\mathscr{L}[1] = \int_{0-}^\infty 1e^{-st}\,dt = \left.\frac{e^{-st}}{-s}\right|_{0-}^\infty = \frac{e^{-s\infty} - e^{-s0}}{-s} = \frac{0 - 1}{-s} = \frac{1}{s}$$

Example 1.10

Find $\mathscr{L}[\delta(t)]$.

Solution

$$\mathscr{L}[\delta(t)] = \int_{0-}^\infty \delta(t)e^{-st}\,dt$$

But because $\delta(t) = 0$ everywhere except at $t = 0$, we only have to integrate from $0-$ to just slightly beyond zero. In this range, $e^{-st} \approx e^0 = 1$, so

$$\mathcal{L}[\delta(t)] = \int_{0-}^{0+} \delta(t)\, dt = 1$$

This is because the integral of a unit impulse is 1.

Example 1.11

Find $\mathcal{L}[t]$.

Solution This is best solved by integration by parts: $\int u\, dv = uv - \int v\, du$. Thus, if $u = t$ and $dv = e^{-st}\, dt$, then $du = dt$ and $v = e^{-st}/-s$.

$$\mathcal{L}[t] = \int_0^\infty te^{-st}\, dt = \left[t\frac{e^{-st}}{-s} \right]_0^\infty - \int_0^\infty \frac{e^{-st}}{-s}\, dt$$

$$= -\frac{t}{se^{+st}}\Bigg|_0^\infty - \frac{e^{-st}}{s^2}\Bigg|_0^\infty$$

$$= -\left(\frac{t}{se^{st}}\right)_{t=\infty} + 0 - \left(\frac{0}{s^2} - \frac{1}{s^2}\right) = -\frac{\infty}{se^{s\infty}} + \frac{1}{s^2}$$

The first term in the last line would seem to be indeterminate since it is ∞/∞, but it can be shown that for any positive s that as $t \rightarrow \infty$, $e^x \rightarrow \infty$ even faster. For instance, with $s = 10$ and $t = 20$, the term $= -0.13867 \times 10^{-85}$, which is close to zero. Or, according to l'Hôpital's rule,

$$\frac{-t}{se^{-st}}\Bigg|^\infty = \frac{\dfrac{d}{dt}(-t)}{\dfrac{d}{dt}(se^{st})}\Bigg|^\infty = \frac{-1}{s^2e^{st}} = \frac{-1}{s^2\infty} = 0$$

Therefore, $\mathcal{L}[t] = 1/s^2$.

Some commonly used transform pairs are listed in Table 1.2. The first three were derived in Examples 1.9 to 1.11. Why is $\mathcal{L}[1]$ the same as $\mathcal{L}[u(t)]$?

Most problems cannot be solved by going directly to the table. Either a constant must be factored out first or the function must be broken down into a sum of two or more functions whose Laplace transform is given in the table. This is the case in the next example.

Example 1.12

Find $\mathcal{L}[i(t)]$ if $i(t) = 5 \cos (6t + 30°)$.

Solution Since this form is not in Table 1.2, we must break this down into the sum of two functions that are in the table. From trigonometry we know that

$$\cos (6t + 30°) = (\cos 6t)(\cos 30°) - (\sin 6t)(\sin 30°)$$

$$= \frac{\sqrt{3}}{2} \cos 6t - \frac{1}{2} \sin 6t$$

TABLE 1.2 COMMON LAPLACE
TRANSFORM PAIRS

$f(t)$	$F(s) = \mathscr{L}[f(t)]$	Pair number
$\delta(t)$	1	(T-1)
1 or $u(t)$	$\dfrac{1}{s}$	(T-2)
t	$\dfrac{1}{s^2}$	(T-3)
$e^{-\alpha t}$	$\dfrac{1}{s + \alpha}$	(T-4)
$\sin \omega t$	$\dfrac{\omega}{s^2 + \omega^2}$	(T-5)
$\cos \omega t$	$\dfrac{s}{s^2 + \omega^2}$	(T-6)
$e^{-\alpha t} \sin \omega t$	$\dfrac{\omega}{(s + \alpha)^2 + \omega^2}$	(T-7)
$e^{-\alpha t} \cos \omega t$	$\dfrac{s + \alpha}{(s + \alpha)^2 + \omega^2}$	(T-8)
t^n	$\dfrac{n!}{s^{n+1}}$	(T-9)
$e^{-\alpha t} t^n$	$\dfrac{n!}{(s + \alpha)^{n+1}}$	(T-10)

Therefore,

$$i(t) = 5 \left(\frac{\sqrt{3}}{2} \cos 6t - \frac{1}{2} \sin 6t \right)$$

Now, using T-5 and T-6, we get

$$\mathscr{L}[i(t)] = 5 \left[\frac{\sqrt{3}}{2} \left(\frac{s}{s^2 + 36} \right) - \frac{1}{2} \left(\frac{6}{s^2 + 36} \right) \right] = 2.5 \left(\frac{\sqrt{3}\, s - 6}{s^2 + 36} \right)$$

1.8 INVERSE LAPLACE TRANSFORM

If $F(s)$ is given and it is desired to find $f(t)$, an inverse procedure is called for. It is symbolized by $\mathscr{L}^{-1}$.

$$\mathscr{L}^{-1}[F(s)] = f(t) \tag{1.10}$$

The formula for the inverse Laplace transform is

$$f(t) = \frac{1}{2\pi j} \int_{\sigma - j\infty}^{\sigma + j\infty} F(s)e^{st}\, ds \qquad (1.11)$$

Notice that this is integration with complex limits. This requires knowledge of a subject called complex variables, so it will not be considered here. Instead, we use one of the following techniques.

The first technique is just to look up the $F(s)$ in the table and read off the corresponding $f(t)$. This will usually involve a preliminary step of factoring out constants until the $F(s)$ looks like one of the forms in Table 1.2. The next three examples demonstrate this technique.

Example 1.13

What is the inverse Laplace transform of $2/(s + 1)$?

Solution   Because of the linearity property, a 2 can be "removed." Thus

$$\mathcal{L}^{-1}\left[\frac{2}{s + 1}\right] = 2\mathcal{L}^{-1}\left[\frac{1}{s + 1}\right] = 2e^{-t} \qquad \text{(by use of pair T-4)}$$

Example 1.14

What is the inverse Laplace transform of $2/(5s + 1)$?

Solution   In this case, a factor of $\frac{2}{5}$ must be "removed." Thus

$$\mathcal{L}^{-1}\left[\frac{2}{5s + 1}\right] = \frac{2}{5}\,\mathcal{L}^{-1}\left[\frac{1}{s + 0.2}\right] = 0.4e^{-t/5} \qquad \text{(by use of pair T-4)}$$

Notice that the "removal" of the 5 from the denominator affected both terms in the denominator.

Example 1.15

Find the inverse Laplace transform of

$$\frac{1}{(2s + 3)^3}$$

Solution   This looks like it might be adaptable to T-10 of Table 1.2 except that this example does not have n (2 in this case) in the numerator, and this example has a 2 as the coefficient of s, not 1. This can all be corrected by using the linearity theorem just by extracting the proper numbers. For example, extract the coefficient of s to the outside of the Laplace operation: it will be $\frac{1}{8}$, since it is cubed. Now place a 2 in the numerator by compensating with $\frac{1}{2}$ on the outside of the Laplace operation. Thus the answer is $\frac{1}{16}$ of the inverse Laplace transform of the standard T-10 form. Thus

$$F(s) = \frac{1}{(2s + 3)^3} = \frac{\frac{1}{2}}{8} \frac{2!}{(s + \frac{3}{2})^3} \rightarrow \frac{1}{16} e^{-(3/2)t}t^2$$

The arrow was used in the last step to show the transformation.

Example 1.16

If

$$F(s) = \frac{2s + 2}{s^2 + 6s + 25}$$

find $f(t)$.

Solution   To see if this will fit T-7 or T-8, it is necessary to examine the coefficients of the terms in the denominator. The coefficient of s^2 in our function as well as in T-7 (or 8) is 1. The coefficient of the s term is 6 and 2α (when T-7 is expanded). Therefore, $\alpha = 3$. Finally, $\alpha^2 + \omega^2$, or $3^2 + \omega^2$ must equal 25. Therefore, $\omega = 4$. Only after thus having analyzed the denominator can we begin to consider the numerator. Since $2s + 2$ is neither 4 or a multiple of 4 nor $s + 3$ or a multiple of $(s + 3)$, our answer is neither T-7 nor T-8, but a combination of the two. Since only T-8 has an s, and our problem has $2s$, we will need twice T-8. Then whatever part of the constant is not used by T-8 must come from T-7. Thus

$$\frac{2s + 2}{s^2 + 6s + 25} = \frac{2(s + 3)}{(s + 3)^2 + 4^2} + \frac{-1(4)}{(s + 3)^2 + 4^2}$$

So $f(t) = 2e^{-3t} \cos 4t - e^{-3t} \sin 4t$.

Another way to find the proper multiples is to assign a letter to each: for instance,

$$2s + 2 = A(s + 3) + B(4)$$

This is an identity (i.e., good for all values of s). Then equating coefficients of s and equating constants, it is found that $A = 2$ and $B = -1$.

1.9 PARTIAL FRACTION EXPANSION OF SOME COMMON FUNCTIONS OF s

Sometimes even with the manipulation shown in Examples 1.14 to 1.16, we will not be able to get the function, $F(s)$, into a form suitable for using Table 1.2. In that case it may be possible to break up the function into a sum of two or more fractions of suitable form. For instance,

$$F(s) = \frac{s + 3}{s^2 + 3s + 2} = \frac{2}{s + 1} + \frac{-1}{s + 2}$$

These individual fractions are of form T-4.

In an algebra course we have learned how to combine two or more fractions into one fraction, but probably never learned to go the other direction. Breaking up one fraction into two or more is called **partial fraction expansion**. But how is this done?

First, it must be realized that the required denominators must be factors of the original denominator.   What is perhaps not so obvious is that if the denominator factors are all different, the numerators will all be constants; call them A, B, C, and so on.

Example 1.17

Break down

$$\frac{s + 3}{s^2 + 3s + 2}$$

into partial fractions.

Solution

$$\frac{s + 3}{s^2 + 3s + 2} = \frac{A}{s + 1} + \frac{B}{s + 2} = \frac{A(s + 2) + B(s + 1)}{s^2 + 3s + 2}$$

$$= \frac{s(A + B) + (2A + B)}{s^2 + 3s + 2}$$

Comparing left and right fractions, the denominators are the same.   Therefore, the numerators are equal to each other:

$$s + 3 = s(A + B) + (2A + B)$$

Since this must be an identity (good for all values of s), the coefficients of s on both sides must be equal and the constants on both sides must be equal.   Therefore,

$$1 = A + B \qquad \text{and} \qquad 3 = 2A + B$$

Solving these two simultaneous equations yields

$$A = 2 \qquad \text{and} \qquad B = -1$$

$$\frac{s + 3}{s^2 + 3s + 2} = \frac{2}{s + 1} + \frac{-1}{s + 2}$$

If the example had asked for the inverse Laplace transform, it would be easy to find it at this point (see T-4 of Table 1.2).   From this example it is seen that the first step is to factor the denominator.   If the denominator is a quadratic and the factors cannot be found by inspection, the quadratic formula can be used.

Example 1.18

Find

$$\mathscr{L}^{-1}\left[\frac{-3}{s^2 + 3s + 1}\right]$$

Solution   Using the quadratic formula gives

$$s = \frac{1}{2}[-3 \pm \sqrt{3^2 - 4(1)(1)}] = -\frac{3}{2} \pm \frac{\sqrt{5}}{2}$$

Therefore,

$$\frac{-3}{s^2 + 3s + 1} = \frac{A}{s + \frac{3}{2} + \sqrt{5}/2} + \frac{B}{s + \frac{3}{2} - \sqrt{5}/2}$$

$$+ \frac{A(s + \frac{3}{2} - \sqrt{5}/2)}{s^2 + 3s + 1} + \frac{B(s + \frac{3}{2} + \sqrt{5}/2)}{s^2 + 3s + 1}$$

Equating like parts of the numerator yields

$$0 = A + B \quad \text{and} \quad -3 = A\left(\frac{3}{2} - \frac{\sqrt{5}}{2}\right) + B\left(\frac{3}{2} + \frac{\sqrt{5}}{2}\right)$$

Solving these two equations simultaneously yields

$$A = 0.6\sqrt{5} \quad \text{and} \quad B = -0.6\sqrt{5}$$

$$\mathcal{L}^{-1}\left[\frac{-3}{s^2 + 3s + 1}\right] = \mathcal{L}^{-1}\left[\frac{0.6\sqrt{5}}{s + \frac{3}{2} + \sqrt{5}/2}\right] + \mathcal{L}^{-1}\left[\frac{-0.6\sqrt{5}}{s + \frac{3}{2} - \sqrt{5}/2}\right]$$

$$\mathcal{L}^{-1}\left[\frac{-3}{s^2 + 3s + 1}\right] = 0.6\sqrt{5}\, e^{-2.618t} - 0.6\sqrt{5}\, e^{-0.382t}$$

$$= 0.6\sqrt{5}\, (e^{-2.618t} - e^{-0.382t})$$

If the quadratic formula had yielded complex or imaginary roots, it means that the denominator could be put into form T-7 or T-8, and the factors would be the conjugate of each other. For instance, if the denominator were $s^2 + 2s + 5$, the factored form would be $(s + 1 + j2)(s + 1 - j2)$. But the use of T-7 and T-8 does not require us to factor but merely to regroup by completing the square. For instance,

$$s^2 + 2s + 5 = (s^2 + 2s + 1) + (4) = (s + 1)^2 + (2)^2$$
$$= (s + \alpha)^2 + \omega^2$$

However, we can use partial fraction expansion to prove T-7 and T-8.

Example 1.19

Prove T-8.

Solution

$$F(s) = \frac{s + \alpha}{(s + \alpha)^2 + \omega^2} = \frac{s + \alpha}{(s + \alpha + j\omega)(s + \alpha - j\omega)}$$

$$= \frac{A}{s + \alpha + j\omega} + \frac{B}{s + \alpha - j\omega} = \frac{A(s + \alpha - j\omega) + B(s + \alpha + j\omega)}{(s + \alpha)^2 + \omega^2}$$

$$= \frac{s(A + B) + A(\alpha - j\omega) + B(\alpha + j\omega)}{(s + \alpha)^2 + \omega^2}$$

Equating powers of *s* and the constants gives

$$A + B = 1 \quad \text{and} \quad A(\alpha - j\omega) + B(\alpha + j\omega) = \alpha$$

which when solved simultaneously gives $A = \frac{1}{2}$ and $B = \frac{1}{2}$. Therefore,

$$\mathcal{L}^{-1}[F(s)] = \mathcal{L}^{-1}\left[\frac{\frac{1}{2}}{s + \alpha + j\omega}\right] + \mathcal{L}^{-1}\left[\frac{\frac{1}{2}}{s + \alpha - j\omega}\right] = \frac{1}{2}\left[e^{(-\alpha - j\omega)t} + e^{(-\alpha + j\omega)t}\right]$$

$$= e^{-\alpha t}\left[\frac{e^{j\omega t} + e^{-j\omega t}}{2}\right] = e^{-\alpha t}\cos \omega t$$

The inverse transform used was T-4. The regrouping in the last line shows the expression in brackets to be a definition of the cosine function (see Section A.1 in Appendix A).

As in the examples, it will generally be true that $F(s)$ will be a fraction whose denominator and numerator are polynomials in *s*. Thus,

$$F(s) = \frac{N(s)}{D(s)} \tag{1.12}$$

where

$$N(s) = a_n(s - z_1)(s - z_2) \cdots (s - z_n) \tag{1.13}$$

and

$$D(s) = b_m(s - p_1)(s - p_2) \cdots (s - p_m) \quad \text{where } m > n \tag{1.14}$$

Cases where $m < n$ will not be considered in this book. If $m = n$, we can divide $N(s)$ by $D(s)$ and obtain a constant (whose inverse is an impulse) plus a new fraction whose numerator degree this time is of lower degree than the denominator degree. If any factor in the numerator is identical to a factor in the denominator, they would cancel, thus simplifying the expression.

Example 1.20

Find the inverse Laplace of

$$\frac{s^2 - 1}{s^2 + 2s + 1}$$

Solution First, we factor this to see if anything cancels.

$$\frac{s^2 - 1}{s^2 + 2s + 1} = \frac{(s + 1)(s - 1)}{(s + 1)(s + 1)} = \frac{s - 1}{s + 1}$$

This has both numerator and denominator of equal degree (*s* is to the first power in both). Either of two methods could be used.

First method: Doing the indicated division yields

$$\frac{s - 1}{s + 1} = 1 + \frac{-2}{s + 1}$$

Inverting both of these terms yields

$$\mathcal{L}^{-1}\left[\frac{s-1}{s+1}\right] = \delta(t) - 2e^{-t}$$

The second method of handling this problem is covered in Example 1.21.

The $z_1, z_2, \ldots$ are the roots of the numerator polynomial and are called the zeros of $F(s)$, because when $s = z_1 \ldots$, $F(s) = 0$. The $p_1, p_2, \ldots$ are the roots of the denominator and are called the poles of $F(s)$, because when $s = p_1 \ldots$, $F(s)$ becomes infinite, and a plot of the function near a pole rises steeply upward and looks like a pole going off the paper. Notice that the roots are subtracted from s when we are talking about factors; if we have seen only plus signs in the factors of our examples, it only means that until now we have dealt only with negative roots.

As we have seen in the examples, the first step in finding $f(t)$ is to factor the denominator of $F(s)$, that is, to find the poles of $F(s)$. We saw that if $m = 2$, we could use the quadratic formula to find the factors. But for higher-degree polynomials we may have to resort to a computer program such as ROOT FINDER. Here, however, we will assume that all denominators that cannot easily be factored will be given to us in factored form, at least into binomial and quadratic factors.

The poles will be either real numbers or complex numbers. Perhaps even some poles will be repeats of others; this is called **multiple-order poles**. In this section we look at just the simple-order real poles. We have already solved functions with two simple poles in Examples 1.17 and 1.18. If we were given a problem with, say, five real poles of simple order, we could solve it by this same method, which in this case would mean solving for five constants (A_1, A_2, A_3, A_4, A_5) from five simultaneous equations. This gets somewhat unwieldy and an alternative method will now be considered.

Consider a function, $F(s)$, that can be written in the form

$$F(s) = \frac{A_1}{s - p_1} + \frac{A_2}{s - p_2} + \cdots + \frac{A_r}{s - p_r} + R(s) \tag{1.15}$$

where p_1 through p_r represent the simple real poles (known because the factors are assumed to be known), A_1 through A_r are unknown at this point, and $R(s)$ is the combination of all the other type factors (multiple real poles, simple complex poles, and multiple complex poles).

If we multiply the equation by the kth factor, $(s - p_k)$, we will obtain

$$(s - p_k)F(s) = A_k + (s - p_k)$$

times all terms on right side, except the kth $\tag{1.16}$

This equation is also an identity, good for all values of s. A particularly convenient value of s would be $s = p_k$. When p_k is substituted for s, the right side of the equation is just A_k. The left side, however, does not equal zero, since the

$(s - p_k)$ factor on the left side immediately cancels the $(s - p_k)$ factor that is contained in the denominator of $F(s)$. Thus

$$A_k = [(s - p_k)F(s)]_{s=p_k} \tag{1.17}$$

This is the formula that we can use to find any of the simple real pole coefficients. The following three examples show this technique.

Another method of handling the case where numerator and denominator degree are equal will now be described. It consists of ignoring this fact for the time being and using the method of partial fraction expansion, as before, followed by Laplace inversion. Then, to this answer, just add $K\delta(t)$, where $K = F(s)|_{s=\infty}$. This is illustrated by the following example.

Example 1.21

Repeat Example 1.16, but using the method described above.

Solution   Ignoring the fact that the degree of numerator and denominator are the same yields

$$F(s), \text{ reduced} = \frac{s - 1}{s + 1} = \frac{A}{s + 1} \quad \text{or} \quad A = \left.\frac{s - 1}{1}\right|_{s=-1} = -2$$

Then

$$\mathcal{L}^{-1}\left[\frac{-2}{s + 1}\right] = -2e^{-t}$$

To this, we add $K\delta(t)$, where

$$K = F(s)|_{s=\infty} = \left.\frac{s - 1}{s + 1}\right|_{s=\infty} = \left.\frac{1 - 1/s}{1 + 1/s}\right|_{s=\infty} = 1$$

So the entire solution is

$$\mathcal{L}^{-1}\left[\frac{s - 1}{s + 1}\right] = \delta(t) - 2e^{-t}$$

which agrees with Example 1.16.

Example 1.22

Repeat Example 1.14.

$$F(s) = \frac{s + 3}{s^2 + 3s + 2} = \frac{A_1}{s + 1} + \frac{A_2}{s + 2}$$

$$A_1 = \left[(s + 1)\frac{s + 3}{s^2 + 3s + 2}\right]_{s=-1} = \left[\frac{s + 3}{s + 2}\right]_{s=-1} = 2$$

$$A_2 = \left[(s + 2)\frac{s + 3}{s + 3s + 2}\right]_{s=-2} = \left[\frac{s + 3}{s + 1}\right]_{s=-2} = -1$$

Therefore,

$$F(s) = \frac{2}{s + 1} + \frac{-1}{s + 2}$$

which agrees with the Example 1.14 result.

Example 1.23

Expand into partial fractions and find $f(t)$ for

$$F(s) = \frac{2s^4 + 27s^3 + 107s^2 + 154s + 76}{(s + 1)(s + 2)(s + 5)(s + 1)}$$

Solution   Since the degree of the numerator and denominator are the same, we expand the denominator and divide out to get the constant.

$$F(s) = \frac{2s^4 + 27s^3 + 107s^2 + 154s + 76}{s^4 + 9s^3 + 25s^2 + 27s + 10} = 2 + \frac{9s^3 + 57s^2 + 100s + 56}{(s + 2) \, (s + 5) \, (s + 1)^2}$$

Since $(s + 1)$ occurs twice, it will be retained as a quadratic for now with a numerator of the form $Bs + C$.

$$F(s) = 2 + \frac{A_1}{s + 2} + \frac{A_2}{s + 5} + \frac{Bs + C}{(s + 1)^2}$$

$$A_1 = \left[2(s + 2) + \frac{9s^3 + 57s^2 + 100s + 56}{(s + 5) \, (s + 1)^2} \right]_{s = -2} = 4$$

$$A_2 = \left[2(s + 5) + \frac{9s^3 + 57s^2 + 100s + 56}{(s + 2) \, (s + 1)^2} \right]_{s = -5} = 3$$

Therefore,

$$F(s) = 2 + \frac{4}{s + 2} + \frac{3}{s + 5} + \frac{Bs + C}{(s + 1)^2}$$

and

$$f(t) = 2\delta(t) + 4e^{-2t} + 3e^{-5t} + \mathcal{L}^{-1}\left[\frac{Bs + C}{(s + 1)^2} \right]$$

1.10 MULTIPLE REAL POLES

If $F(s)$ contains poles of multiplicity r, then $F(s)$ may be written as

$$F(s) = \frac{Q(s)}{(s - p)^r} \tag{1.18}$$

where p is the pole of order r and

$$Q(s) = F(s) \, (s - p)^r \tag{1.19}$$

It can be shown that $F(s)$ can also be written in the form

$$F(s) = \frac{B_1}{(s - p)^r} + \frac{B_2}{(s - p)^{r-1}} + \cdots + \frac{B_r}{s - p} + R(s) \qquad (1.20)$$

where the unknown B coefficients are due to the multiple pole and $R(s)$ is the expansion due to all the other poles [$R(s)$ is not the same $R(s)$ mentioned in Eq. 1.15]. Finally, it can be shown that

$$B_k = \frac{1}{(k - 1)!} \frac{d^{k-1}}{ds^{k-1}} Q(s) \bigg|_{s=p} \qquad (1.21)$$

This formula can be used to find any of the B coefficients. It is important that the subscripts in Eq. 1.21 match those established in Eq. 1.20. If $k = 5$, for instance, then Eq. 1.21 is

$$B_5 = \frac{1}{4!} \frac{d^4}{ds^4} Q(s) \bigg|_{s=p}$$

which is $\frac{1}{24}$ of the fourth derivative of $Q(s)$, which is then evaluated at $s = p$. It is important that the evaluation be done after the derivative has been found.

Example 1.24

Finish Example 1.23.

Solution   By Eq. 1.19,

$$Q(s) = \frac{2s^4 + 27s^3 + 107s^2 + 154s + 76}{(s + 2)(s + 5)}$$

$$= (2s^4 + 27s^3 + 107s^2 + 154s + 76)(s^2 + 7s + 10)^{-1}$$

From Eq. 1.21,

$$B_2 = \frac{1}{1!} [-1(s^2 + 7s + 10)^{-2}(2s + 7)(2s^4 + 27s^3 + 107s^2 + 154s + 76)$$

$$+ (s^2 + 7s + 10)^{-1}(8s^3 + 81s^2 + 214s + 154)]_{s=-1} = 2$$

Also,

$$B_1 = Q(s)|_{s=-1} = 1$$

$$F(s) = 2 + \frac{4}{s + 2} + \frac{3}{s + 5} + \frac{1}{(s + 1)^2} + \frac{2}{s + 1}$$

Using the results of Example 1.23 and T-10 and T-4 for the other two gives

$$f(t) = 2\delta(t) + 4e^{-2t} + 3e^{-5t} + te^{-t} + 2e^{-t}$$

1.11 COMPLEX POLES OF SIMPLE ORDER

If complex poles occur, they will in general yield damped sine waves, damped cosine waves, or some combination of the two. This is so because the complex poles produce denominators such as are in T-7 or T-8. The numerator is of the form $Bs + C$, but B and C are unknown. If they were known, we would be able to solve the problem as in Example 1.16. Three methods will be used to find the time function due to the complex poles.

     A. Expansion as a Quadratic. The first method is to find B and C and then use T-7 and T-8. First we must do as much expansion (the real poles) as possible and then substitute values of s into this partially expanded $F(s)$ in order to generate two simultaneous equations. The values we use should be easy numbers such as 0, 1, -1, and so on, but cannot be equal to any poles we have already found.

Example 1.25

Expand

$$F(s) = \frac{5s + 7}{(s + 1)(s^2 + 2s + 5)}$$

and find $\mathscr{L}^{-1}[F(s)]$.

Solution   Since there are one real and two complex poles, $F(s)$ is written as

$$F(s) = \frac{A}{s + 1} + \frac{Bs + C}{s^2 + 2s + 5}$$

Now find the A (for the real pole at $s = -1$) by using Eq. 1.17:

$$A = \frac{5s + 7}{s^2 + 2s + 5}\bigg|_{s = -1} = \frac{2}{4} = \frac{1}{2}$$

Therefore,

$$F(s) = \frac{\frac{1}{2}}{s + 1} + \frac{Bs + C}{s^2 + 2s + 5}$$

For s, substitute first 0, then 1:

$$\frac{7}{1(5)} = \frac{\frac{1}{2}}{1} + \frac{C}{5} \qquad\qquad \text{from which } C = \tfrac{9}{2}$$

and

$$\frac{5 + 7}{2(1 + 2 + 5)} = \frac{\frac{1}{2}}{1 + 1} + \frac{B + \frac{9}{2}}{1 + 2 + 5} \qquad \text{from which } B = -\tfrac{1}{2}$$

Therefore,

$$F(s) = \frac{\frac{1}{2}}{s + 1} + \frac{-s/2 + \frac{9}{2}}{s^2 + 2s + 5} = \frac{1}{2}\frac{1}{s + 1} + \frac{-\frac{1}{2}(s + 1) + \frac{5}{2}(2)}{(s + 1)^2 + 2^2}$$

Notice that the numerator of the last fraction has been rearranged so that T-7 and T-8 can be used. Inverting, we get

$$f(t) = \tfrac{1}{2}e^{-t} - \tfrac{1}{2}e^{-t} \cos 2t + \tfrac{5}{2}e^{-t} \sin 2t$$

$$= \tfrac{1}{2}e^{-t}[1 + (5 \sin 2t - \cos 2t)] = \tfrac{1}{2}e^{-t}[1 + 5 \sin 2t - \sin (2t + 90°)]$$

$$= \tfrac{1}{2}e^{-t}[1 + \sqrt{26} \sin (2t - \tan^{-1} (-\tfrac{1}{5}))] \qquad \text{by adding the sine phasors}$$

$$= \tfrac{1}{2}e^{-t}[1 + 5.1 \sin (2t - 11.3°)]$$

B. Expansion by Poles.   The second method does not require us to expand the real poles first. We can expand the complex poles directly as we did for real poles, but this involves dealing with complex numbers, which at times can get somewhat confusing, but we will try it anyway.

Example 1.26

Find the portion of the expansion of

$$F(s) = \frac{5s + 7}{(s + 1)(s^2 + 2s + 5)}$$

due to the complex poles.

Solution   We have seen that this portion is

$$\frac{Bs + C}{s^2 + 2s + 5}$$

but now we must expand in terms of the individual complex poles. These are

$$\frac{K_1}{s + 1 + j2} + \frac{K_2}{s + 1 - j2}$$

Using Eq. 1.17 gives us

$$K_1 = \frac{5s + 7}{(s + 1)(s + 1 - j2)}\bigg|_{s = -1 - j2} = \frac{-5 - j10 + 7}{(-1 - j2 + 1)(-1 - j2 + 1 - j2)}$$

$$= \frac{2 - j10}{(-j2)(-j4)} = -\frac{1}{4} + j\frac{5}{4}$$

$$K_2 = \frac{5s + 7}{(s + 1)(s + 1 + j2)}\bigg|_{s = -1 + j2} = \frac{-5 + j10 + 7}{(-1 + j2 + 1)(-1 + j2 + 1 + j2)}$$

$$= \frac{2 + j10}{(j2)(j4)} = -\frac{1}{4} - j\frac{5}{4}$$

Thus K_1 and K_2 are complex conjugates of each other. This will always be the case as long as the coefficients in the quadratic denominator are real. So once we find K_1, we do not have to go through the math to find K_2; just take the conjugate of K_1.

Therefore,

$$\frac{Bs + C}{s^2 + 2s + 5} = \frac{-\frac{1}{4} + j\frac{5}{4}}{s + 1 + j2} + \frac{-\frac{1}{4} - j\frac{5}{4}}{s + 1 - j2} = \frac{-s/2 + \frac{9}{2}}{s^2 + 2s + 5}$$

from which $B = -\frac{1}{2}$ and $C = \frac{9}{2}$, agreeing with Example 1.25.

If the B and C form is used, then T-7 and T-8 will be used to find the inverse Laplace transform, as in Example 1.16. If the K_1 and K_2 form is used, T-4 can be used on each fraction to find the inverse. This will yield complex exponentials which can then be combined to get sine and cosine terms as was done in Example 1.19.

C. Inverse Transform Using the "Trick" Method. The third method, called the "trick" formula, will lead to the time function directly. The advantage is that this is a very quick process. The disadvantage is that we never get to see the expanded version of $F(s)$.

The first step in this method is to find $Q(s)$, which is defined as $F(s)$ with the complex-pole denominator factor removed. Thus

$$Q(s) = F(s) \left[(s + \alpha)^2 + \omega^2 \right] \tag{1.22}$$

In this form the poles are identified as $(-\alpha + j\omega)$ and $(-\alpha - j\omega)$. The second step is to evaluate $Q(s)$ at $s = -\alpha + j\omega$ and convert it to polar form.

$$Q(-\alpha + j\omega) = M \angle \theta \tag{1.23}$$

The third step is to write the answer using the values above in the following "trick" formula:

$$f(t) = \frac{M}{\omega} e^{-\alpha t} \sin (\omega t + \theta) \tag{1.24}$$

Equation 1.24 is a damped sinusoid, but if α would be zero, $f(t)$ would be an undamped sinusoid.

Example 1.27

Repeat Example 1.26 using the "trick" formula.

Solution   Applying Eq. 1.22 yields

$$Q(s) = \frac{5s + 7}{s + 1}$$

Using the quadratic formula on $(s^2 + 2s + 5)$, we find that the root with the positive j term is $-1 + j2$.

$$Q(-1 + j2) = \frac{5(-1 + j2) + 7}{(-1 + j2) + 1} = 5 - j = 5.1 \angle -11.3°$$

$$f(t) = \frac{5.1}{2} e^{-t} \sin (2t - 11.3°)$$

which agrees with Example 1.25. Thus the third method seems to be the quickest method.

1.12 MULTIPLE-ORDER COMPLEX POLES

Here we must use the second method of the preceding section, after first factoring the multiple quadratic factors. For example, if the multiple complex factor is $[s^2 + 2\alpha s + (\alpha^2 + \omega^2)]^3$, it would factor into $(s + \alpha + j\omega)^3 (s + \alpha - j\omega)^3$. Then each one of these triple poles could be handled by the second method of the preceding section. However, after finding the three numerator constants for the first triple root, the three for the second triple root are just the complex conjugates of the first three.

Example 1.28

Find

$$\mathcal{L}^{-1}\left[\frac{1}{(s^2 + 2s + 10)^2}\right]$$

Solution

$$\frac{1}{(s^2 + 2s + 10)^2} = \frac{1}{[(s + 1)^2 + 3^2]^2} = \frac{1}{[(s + 1 + j3)(s + 1 - j3)]^2}$$

$$= \frac{1}{(s + 1 + j3)^2 (s + 1 - j3)^2}$$

Notice the regrouping that was done in the last line. This reduces the problem to two sets of multiple roots. Since the first set is a double pole of $-1 - j3$, the $Q(s)$ for this set is

$$Q(s) = F(s) [(s + 1 + j3)^2] = \frac{1}{(s + 1 - j3)^2}$$

Therefore,

$$A_1 = \left.\frac{1}{(s + 1 - j3)^2}\right|_{s = -1 - j3} = \frac{1}{(-1 - j3 + 1 - j3)^2} = -\frac{1}{36}$$

$$A_2 = \left.\frac{1}{(2 - 1)!}\frac{d}{ds}\left[\frac{1}{(s + 1 - j3)^2}\right]\right|_{s = -1 - j3} = \left.\frac{-2(1)}{(s + 1 - j3)^3}\right|_{s = -1 - j3}$$

$$= \frac{-2}{(-1 - j3 + 1 - j3)^3} = \frac{-2}{(-j6)^3} = +\frac{j}{108}$$

It can be shown that if any denominator factors are complex conjugates of each other, their respective numerators after expansion are also complex conjugates of each other. In this example, that would mean that

$$B_1 = -\frac{1}{36} \quad \text{and} \quad B_2 = \frac{-j}{108}$$

Collecting these results and factoring out 108, we get

$$\frac{1}{(s^2 + 2s + 10)^2} = \frac{1}{108}\left[\frac{-3}{(s + 1 + j3)^2} + \frac{j}{s + 1 + j3} + \frac{-3}{(s + 1 - j3)^2}\right.$$

$$\left. + \frac{-j}{s + 1 - j3}\right]$$

Using T-4 for the second and fourth terms and T-10 for first and third terms gives

$$F(s) = \frac{1}{108}\left[-3te^{-(1+j3)t} + je^{-(1+j3)t} - 3te^{-(1-j3)t} - je^{-(1-j3)t}\right]$$

$$= \frac{-1}{18}te^{-t}\frac{e^{j3t} + e^{j3t}}{2} + \frac{1}{54}e^{-t}\frac{e^{j3t} - e^{-j3t}}{2j}$$

The expressions in parentheses can be converted into sine and cosine functions

$$F(s) = -\tfrac{1}{18}\,te^{-t}\cos 3t + \tfrac{1}{54}\,e^{-t}\sin 3t$$

1.13 COMPUTER-AIDED PARTIAL FRACTION EXPANSION

Many times, the $F(s)$ will not have its denominator in factored form. If it is of third degree or higher, it may be difficult to factor. In that case, the program called ROOT FINDER, presented in Appendix B, can be useful.

This program will factor any polynomial in s into linear factors, by giving you the roots of that polynomial. The program first asks for the degree, N. This is the highest power of s in the unfactored polynomial. Then it asks for the coefficients of each power of s. It assumes that these coefficients might be complex, but in our problems all our coefficients are real. This means that you are to answer zero when the imaginary part is asked for. The program needs the coefficients starting from the zeroth power of s (the constant term). If any power is missing, you must reply with zero real and zero imaginary.

When all the data are entered, the computer displays the roots. Each root has two parts: the real in the left column and the imaginary in the right column. Because the program gives only approximate results, you must use your judgment in rounding off the results, or even letting part or all of the root equal zero. Finally, we must construct the factors in the form $(s - \text{root})$. The following example will illustrate these points.

Example 1.29

Factor $s^5 + 10s^4 + 46s^3 + 114s^2 + 117s$.

Solution After entering zero and zero for the first coefficient (since there is no constant term), the other coefficients are entered and the following real and imaginary numbers are displayed:

Real	Imaginary	Interpretation		Factor
−2.22044605E-16	−8.55426841E-14	0	0	s
−2.99993043	1.78598353E-05	−3	0	$s + 3$
−3.00006957	−1.78593381E-05	−3	0	$s + 3$
−2	3	−2	3	$s + 2 - 3j$
−2	−3	−2	−3	$s + 2 + 3j$

Notice that the root changes sign when put into the factor form. Hence

$$s^5 + 10s^4 + 46s^3 + 114s^2 + 117s = s(s + 3)^2 (s + 2 - 3j) (s + 2 + 3j)$$

You could combine the last two factors, yielding the following factored form: $s(s + 3)^2 (s^2 + 4s + 13)$.

If it is required to expand $F(s)$ by partial fractions, it will be necessary to have the denominator in factored form, which can be done as in the preceding discussion, but of course it is not necessary to factor the numerator. It is necessary, however, to go through the procedure of Sections 1.9 and 1.10 in order to find the individual numerators of the partial fractions. These numerator constants are called "residues" by some authors. It is possible to use the computer program called PARTIAL FRACTION EXPANSION given in Appendix B to find these numerator constants, provided that we already have the denominator factored into linear (no quadratic factors) factors.

The program first asks for the degree of the denominator (the highest power of s in the unfactored denominator). Then it asks for the numerator coefficients starting with the constant term, but this time it only takes a real value; that is, if the numerator were to contain imaginary numbers, this program would not work. When entering data, remember to enter zero for the coefficient of any missing power of s. Then it asks for more information about the denominator again. This time it wants the roots (not the coefficients) of the denominator. These are the roots found by running the ROOT FINDER program as described previously. These must be entered in a certain order. Start with the root having the lowest real part: for instance, the roots $-2 + j1$, $-2 - j1$, $-1 - j8$, $-1 + j8$, $0 + j0$, $+1$, $+1$, $+6$ are in the proper order. It is extremely important that you remember in which order you entered these roots, since the output of the program will be in the same order; this will help you get the right numerator constants above the right denominator factors. Double roots are entered twice, which means that the first "residue" for the double goes over the square factor. An example follows.

Example 1.30

Perform PFE on

$$F(s) = \frac{2s^4 + 5s^2 + 11s + 12}{s^5 + 10s^4 + 46s^3 + 114s^2 + 117s}$$

Solution   The first step is to factor the denominator (see Example 1.29). It is now best to write the partial fractions with the numerators missing, so as to get the fractions set up in ascending order of the real part of the roots. This is

$$F(s) = \frac{}{(s + 3)^2} + \frac{}{s + 3} + \frac{}{s + 2 - 3j} + \frac{}{s + 2 + 3j} + \frac{}{s}$$

Now load the PFE program. Enter 5 for the denominator degree. Then enter the numerator coefficients:

12, return, 11, return, 5, return, 0, return, 2, return

Now enter the real and imaginary parts of the denominator roots. These should be in the order shown in the equation above, except that the signs are reversed, since you are entering roots, not factors:

$$-3, 0; \quad -3, 0; \quad -2, 3; \quad -2, -3; \quad 0, 0$$

When the program sees the double entry of $-3,0$, it realizes that this is a double root and will display the residues in the proper order; that is, the first number will be the one that belongs above the square term, and the second number will belong above the single $(s + 3)$. So these results can just be copied onto the blank spaces in the equation above. You should not change the sign of these results, since they are not roots or factors; they are merely numerator constants. Thus the answer is

$$F(s) = \frac{-6.2}{(s + 3)^2} + \frac{4.527}{s + 3} + \frac{-1.315 + j0.913}{s + 2 - 3j} + \frac{-1.315 - j0.913}{s + 2 + 3j} + \frac{0.1025}{s}$$

To take the inverse Laplace transform of this expression, it would be more convenient if the two complex fractions were combined into one; then formulas T-7 and T-8 could be used to find that portion of the answer. Unfortunately, the computer program will not supply an output in this form. That means that we must use it as it is and use transform T-4 together with a great deal of algebraic manipulation and the use of formulas A1.5 and A1.6 of Appendix A, or we must combine these two fractions ourselves by putting them over a common denominator, and then breaking the new numerator up into two parts so as to obtain the standard transform pairs of T-7 and T-8. The other fractions can, of course, be easily transformed into the time domain by using T-4 and T-10 and T-2.

1.14 LAPLACE TRANSFORM THEOREMS

In addition to the properties shown in Eq. 1.8 and 1.9, there are five other very useful operations or theorems involving Laplace transforms. They are summarized in Table 1.3. What this table says is that if we know $f(t)$ or its transform $F(s)$, but we want to know the transform of $f'(t)$ or of the integral of $f(t)$ or of $f(t)$ multiplied by a damping factor, it will be very simple to do this.

These theorems will be illustrated by the following examples. Some of these offer an alternative way of proving some of the transform pairs in Table 1.2 in combination with one of the theorems from Table 1.3.

Example 1.31

If $f(t) = t^n$, find $\mathcal{L}[f'(t)]$, using O-1.

Solution   We know from Table 1.2 that $F(s) = n!/s^{n+1}$. Also, $f(0) = 0^n = 0$. Therefore,

$$f'(t) = \frac{sn!}{s^{n+1}} - 0 = \frac{n!}{s^n}$$

We will check this by actually differentiating and then taking the Laplace transform.

$$f'(t) = n(t^{n-1})$$

$$\mathcal{L}[f'(t)] = n\mathcal{L}[t^{n-1}] = n\frac{(n-1)!}{s^{(n-1)+1}} = \frac{n!}{s^n} \qquad \text{checks}$$

Example 1.32

If $f(t) = e^{-\alpha t}$, find $\mathcal{L}[f'(t)]$, using O-1.

Solution

$$\mathcal{L}[e^{-\alpha t}] = \frac{1}{s+\alpha} \qquad \text{and} \qquad f(0) = e^{-\alpha 0} = 1$$

Therefore,

$$\mathcal{L}[f'(t)] = \frac{s}{s+\alpha} - 1 = -\frac{\alpha}{s+\alpha}$$

TABLE 1.3 LAPLACE THEOREMS

Type of time operation	If $\mathcal{L}[f(t)] = F(s)$, then [for same $f(t)$ and $F(s)$]:		Formula number
Differentiation	$\mathcal{L}[f'(t)]$	$= sF(s) - f(0)$	(O-1)
Integration	$\mathcal{L}\left[\int_0^t f(t)dt\right]$	$= \dfrac{F(s)}{s}$	(O-2)
Time shift[a]	$\mathcal{L}[f(t-a)u(t-a)]$	$= e^{-as}F(s)$	(O-3)
Damping	$\mathcal{L}[e^{-\alpha t}f(t)]$	$= F(s+\alpha)$	(O-4)
Time scaling	$\mathcal{L}[f(at)]$	$= \dfrac{1}{a}F\left(\dfrac{s}{a}\right)$	(O-5)

[a]The amount of the time shift, a, must be positive.

If we actually differentiate $f(t)$, we get $-\alpha e^{-\alpha t}$, whose transform is

$$-\alpha \frac{1}{s+\alpha}$$

so it checks again.

Example 1.33

Find $\mathcal{L}\left[\int_0^t \cos \omega t \, dt\right]$, using O-2.

Solution

$$\mathcal{L}\left[\int_0^t \cos \omega t \, dt\right] = \frac{1}{s}\mathcal{L}[\cos \omega t] = \frac{1}{s}\frac{s}{s^2+\omega^2} = \frac{1}{s^2+\omega^2}$$

This can be checked by actually integrating:

$$\int_0^t \cos \omega t \, dt = \frac{1}{\omega}\sin \omega t \, \bigg|_0^t = \frac{1}{\omega}\sin \omega t - 0$$

whose transform is

$$\frac{1}{\omega}\frac{\omega}{s^2+\omega^2} = \frac{1}{s^2+\omega^2}$$

which agrees with the above.

Example 1.34

Find $\mathcal{L}^{-1}\left[e^{-2s}\frac{2s+2}{s^3}\right]$.

Solution   First we must find

$$\mathcal{L}^{-1}\left[\frac{2s+2}{s^3}\right] = \mathcal{L}^{-1}\left[\frac{2}{s^2}\right] + \mathcal{L}^{-1}\left[\frac{2}{s^3}\right] = 2t + t^2$$

Then by O-3,

$$\mathcal{L}^{-1}\left[e^{-2s}\frac{2s+2}{s}\right] = [2(t-2) + (t-2)^2]u(t-2)$$

$$= [t^2 - 2t]u(t-2)$$

Example 1.35

Find the Laplace transform of the square wave of Figure 1.4.

Solution Since $f(t) = u(t) - u(t - 1) + u(t - 2) - \cdots$, and $\mathscr{L}[u(t)] = 1/s$, then using O-3,

$$\mathscr{L}[u(t - 1)] = \frac{e^{-s}}{s},$$

$$\mathscr{L}[u(t - 2)] = \frac{e^{-2s}}{s}, \qquad \text{etc.}$$

Therefore,

$$\mathscr{L}[\text{square wave}] = \frac{1}{s} - \frac{e^{-s}}{s} + \frac{e^{-2s}}{s} - \cdots$$

$$= \frac{1}{s}(1 - e^{-s} + e^{-2s} - \cdots)$$

$$= \frac{1}{s}\frac{1}{1 + e^{-s}} \qquad \text{(closed form)}$$

Example 1.36

Derive T-7 from T-5 and T-10 from T-9.

Solution Since the time function of T-7 is just that of T-5 multiplied by $e^{-\alpha t}$, then by O-4 we just replace s in T-5 with $(s + \alpha)$, yielding T-7 directly. The same argument holds for finding T-10 from T-9.

Example 1.37

Using O-5 and knowing that $\mathscr{L}[e^{-t}] = 1/(s + 1)$, find $\mathscr{L}[e^{-\alpha t}]$.

Solution Since the new function of time is just the old one with αt replacing t, then O-5 can be applied. Therefore, the new s function is

$$\frac{1}{\alpha}\frac{1}{(s/\alpha) + 1} = \frac{1}{s + \alpha}$$

1.15 INITIAL AND FINAL VALUE THEOREMS

Occasionally, we encounter a known $F(s)$ whose $f(t)$ is difficult to find. In some of these cases, perhaps we are only interested in the value of $f(t)$ at $t = 0$ or perhaps after the function has stabilized (at t = approximately infinity). In these cases there are methods called the initial value theorem and the final value theorem that will give us these values of $f(t)$ without actually having to find the inverse Laplace transform and substituting $t = 0$ or infinity. These are*

$$f(0+) = \lim_{s \to \infty} sF(s) \tag{1.25}$$

$$f(\infty) = \lim_{s \to 0} sF(s) \tag{1.26}$$

Notice that $F(s)$ must be multiplied first by s, and then the proper value of s substituted. The value of s used is just the opposite of what you might think. For $f(0+)$ you must substitute ∞ for s. For $f(\infty)$ you must substitute 0 for s. For instance, if $F(s) = 1/s^2$, then

$$f(\infty) = \left.\frac{s}{s^2}\right|_{s=0} = \left.\frac{1}{s}\right|_{s=0} = \frac{1}{0} = \infty$$

To check this we know that if $F(s) = 1/s^2$, then $f(t) = t$. Thus $f(\infty)$ does equal infinity.

In many of these cases of substitution of the limits, we will get $0/0$ or ∞/∞. To handle this, we must try to simplify as much as possible to avoid this condition. The next example should illustrate this.

Example 1.38

Find $f(0+)$ and $f(\infty)$ if

$$F(s) = \frac{5s^3 + 3s}{s^2(s + 3)(s + 4)}$$

Solution

$$f(0+) = sF(s)\bigg|_{s=\infty} = \frac{5s^4 + 3s^2}{s^2(s + 3)(s + 4)} = \frac{\infty}{\infty}$$

To avoid this, we first divide top and bottom by s^4:

$$f(0+) = \frac{5 + 3/s^2}{1(1 + 3/s)(1 + 4/s)}\bigg|_{s=\infty} = \frac{5 + 0}{1(1 + 0)(1 + 0)} = 5$$

Another way is to think of $3s^2$ as being small compared to $5s^4$ when $s \to \infty$, so that the $3s^2$ term can be ignored. By the same reasoning, $s + 3$ is approximately $= s$ and $s + 4$ is also approximately $= s$. Therefore, $sF(s) \approx 5s^4/s^4 = 5$ when s is large.

Using Eq. 1.26 to find $f(\infty)$, we get $0/0$. To avoid this difficulty, we can divide top and bottom by s^2 in such a way as to avoid these zeros:

$$sF(s)\bigg|_{s=0} = \frac{5s^2 + 3}{1(s + 3)(s + 4)}\bigg|_{s=0} = \frac{3}{12} = \frac{1}{4}$$

*Equation 1.26 applies only if $sF(s)$ has all its poles in the left half-plane; that is, the poles of $sF(s)$ must all be of the form: a or $a + jb$ (a is negative).

Another way is to think that as $s \to 0$, $s + 3 \to 3$ and $s + 4 \to 4$, and the s^2 term in the denominator divides the numerator to $5s^2 + 3$ and so $5s^2 + 3 \to 3$. With these approximations, which are good as $s \to 0$, then

$$f(\infty) = sF(s) \bigg|_{s=0} = \frac{3}{3(4)} = \frac{1}{4}$$

Example 1.39

Find $f(0+)$ and $f(\infty)$ if

$$F(s) = \frac{s^3 + 3s}{s + 1}$$

Solution

$$f(0+) = s \frac{s^3 + 3s}{s + 1} \bigg|_{s=\infty} = \frac{\infty}{\infty} \quad \text{or better} \quad = \frac{s^3 + 3s}{1 + 1/s} \bigg|_{s=\infty} = \frac{\infty}{1} = \infty$$

$$f(\infty) = s \frac{s^3 + 3s}{s + 1} \bigg|_{s=0} = \frac{0(0)}{0 + 1} = \frac{0}{1} = 0$$

SUMMARY

Processes and phenomena in many branches of science can be described by means of ordinary differential equations. The solution of these equations is necessary to describe the explicit relationship between the variables. The method of Laplace transforms simplifies the solution of these equations by transforming the problem into one that is essentially algebraic in nature. In this chapter we explain the techniques involved in doing this transformation.

Typical functions occurring in linear, passive, time-invariant systems are usually combinations of either impulse or step functions and the more familiar mathematical functions. It is then shown how to find the Laplace transform of these functions and, most important, how to find the inverse of the Laplace transform. Partial fraction expansion plays an important role in conditioning the transform so that this transform can easily be inverted. Computer programs are available which can find factors of polynomials and even calculate the numerators of these partial fractions.

Laplace theorems are helpful in finding the transforms of functions that are related to other functions whose transforms are known. The initial and final value theorems are a quick way to find the initial or final value of a time function when only its transform is known and when it is desired to avoid the work of finding its inverse.

PROBLEMS

Sections 1.1 and 1.2

1.1. Express the functions of Figure P1.1 in terms of step functions.

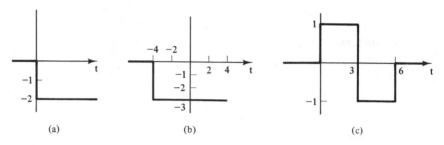

(a)   (b)   (c)

Figure P1.1

1.2. Express the function of Figure P1.2 in terms of step functions.

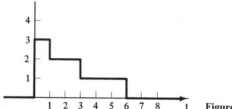

Figure P1.2

1.3. Express the four cycles of Figure P1.3 in terms of step functions.

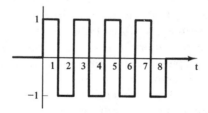

Figure P1.3

1.4. Express the four cycles of Figure P1.4 in terms of step functions.

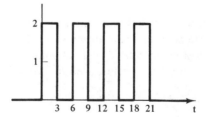

Figure P1.4

1.5. Sketch the following three functions.
   (a) $u(5 - t)$ **(b)** $-3u(4 - t)$ **(c)** $-3u(t - 4)$
1.6. Sketch the following two functions.
   (a) $u(3 - t) + u(t - 3)$ **(b)** $u(t) + u(t + 1) + u(t - 1) - 3u(t - 2)$
1.7. In Figure P1.7 the switch begins at position 1 and rotates clockwise, pausing 1 second at each position (at $t = 0$, the switch has just arrived at position 1). Sketch v_{out} for $t > 0$.

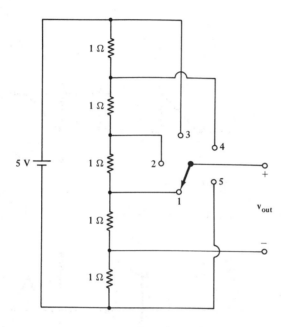

Figure P1.7

Sections 1.3 and 1.4

1.8. Sketch the output, $v_c(t)$, for the circuit of Figure P1.8. Assume that the initial charge on the capacitor is zero. Give the proper mathematical expression for $v_c(t)$.

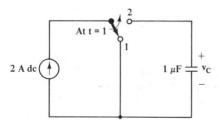

Figure P1.8

1.9. Find expressions for the four plots of Figure P1.9.

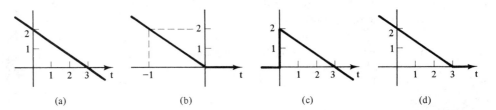

(a)   (b)   (c)   (d)

Figure P1.9

1.10. Find expressions for the three plots of Figure P1.10.

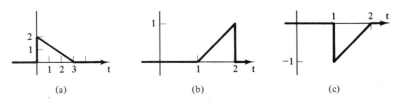

(a)   (b)   (c)

Figure P1.10

1.11. Find expressions for the four plots of Figure P1.11.

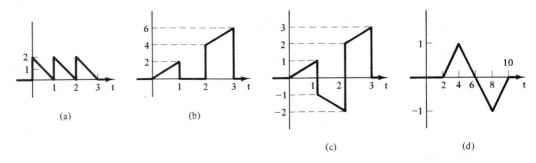

(a)   (b)

(c)   (d)

Figure P1.11

1.12. Plot the following two functions.

(a) $tu(t) - (t - 1)u(t - 1)$ (b) $tu(t) - 2(t - 1)u(t - 1)$

1.13. Plot $f(t) = tu(t) - (t - 1)u(t - 1) + (t - 2)u(t - 2)$

1.14. Find an expression for the plot of Figure P1.14.

Figure P1.14

1.15. Find an expression for the plot of Figure P1.15.

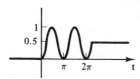

Figure P1.15

Section 1.5

1.16. If Figure P1.4 represents the voltage across a 1-μF capacitor, plot the capacitor current.

1.17. Plot $\delta(t - 1) \sin (\pi t/6)$ and find its simplified expression.

Sections 1.6 and 1.7

1.18. Find the Laplace transform for the following equations.
(a) $q(t) = 20$ (b) $i(t) = 2 + 3t$ (c) $v(t) = 5t(2 + 3t)$

1.19. By the use of Eq. 1.6, find the Laplace transform of the following equations.
(a) $x(t) = \delta(t - 2)$ (b) $y(t) = t^2\delta(t - 2)$ (c) $z(t) = 5u(t - 3)$

1.20. What is the Laplace transform of $f(t) = 5(t - 3)e^{-2t}$?

1.21. Derive T-4 of Table 1.2 by using Eq. 1.6.

1.22. Derive T-5 of Table 1.2 by using Eq. 1.6. (*Hint:* Integrate by parts twice.)

1.23. Find $\mathcal{L}[100 \sin (50t - 45°)]$.

1.24. Using the Laplace definition (Eq. 1.6), find $\mathcal{L}[tu(t - a)]$.

1.25. Find the Laplace transform of the following functions.
(a) $2t^2 - e^{-t}$ (b) $3t^4 - 2t^3 + 4e^{-3t} - 2 \sin 5t + 3 \cos 2t$
(c) $(t^2 + 1)^2$ (d) $(t + 2)^2e^t$

1.26. Find the Laplace transform of the following functions.
(a) $2e^{4t}$ (b) $5t - 3$ (c) $3 \cos 5t$
(d) $10 \sin 6t$ (e) $6 \sin 2t - 5 \cos 2t$

1.27. Find the Laplace transform of the following functions.
(a) t^3e^{-3t} (b) $e^{-t} \cos 2t$ (c) $2e^{3t} \sin 4t$ (d) $e^{2t}(3 \sin 4t - 4 \cos 4t)$

Section 1.8

Find the inverse Laplace transforms.

1.28. (a) $\dfrac{3}{s + 4}$ (b) $\dfrac{1}{2s - 5}$ (c) $\dfrac{8s}{s^2 + 16}$

(d) $\dfrac{6}{s^2 + 4}$ (e) $\dfrac{3s - 12}{s^2 + 18}$

1.29. (a) $\dfrac{1}{s^5}$ (b) $\dfrac{12}{4 - 3s}$ (c) $\dfrac{3s - 8}{4s^2 + 25}$

(d) $\dfrac{3s + 2}{4s^2 + 12s + 9}$ (e) $\dfrac{4s + 12}{s^2 + 8s + 16}$

Section 1.9

Find the inverse Laplace transforms.

1.30. (a) $\dfrac{2s + 1}{s(s + 1)}$ **(b)** $\dfrac{5s + 10}{9s^2 - 16}$ **(c)** $\dfrac{3s - 14}{s^2 - 4s + 8}$ **(d)** $\dfrac{8s + 20}{s^2 - 12s + 32}$

1.31. (a) $\dfrac{2s^2 - 4}{(s + 1)(s - 2)(s - 3)}$ **(b)** $\dfrac{3s + 7}{s^2 - 2s - 3}$ **(c)** $\dfrac{6s - 4}{s^2 - 4s + 20}$

1.32. (a) $\dfrac{s + 5}{s + 4}$ **(b)** $\dfrac{s^3 + 3s^2 + 2s}{2s^3 + 10s^2 + 12s}$

Section 1.10

Find the inverse Laplace transforms.

1.33. (a) $\dfrac{1}{s^3(s + 1)}$ **(b)** $\dfrac{s + 2}{s^4(s + 3)}$ **(c)** $\dfrac{1}{s(s + 1)^3}$

1.34. (a) $\dfrac{1}{(s - 1)^5(s + 2)}$ **(b)** $\dfrac{s}{(s - 2)^5(s + 1)}$

Sections 1.11 and 1.12

1.35. Find the inverse Laplace transforms.

(a) $\dfrac{3s + 1}{(s - 1)(s^2 + 1)}$ **(b)** $\dfrac{2s^3 + 10s^2 + 8s + 40}{s^2(s^2 + 9)}$ **(c)** $\dfrac{s + 1}{(s^2 + 2s + 2)^2}$

Section 1.13

1.36. Use the ROOT FINDER program in Appendix B to factor the polynomial $2s^3 - 4s^2 - 46s + 120$.

1.37. Expand the $F(s)$ of Problem 1.35a into partial fractions by using the PARTIAL FRACTION EXPANSION program in Appendix B.

Section 1.14

1.38. If $\mathcal{L}[\sin 5t] = 5/(s^2 + 25)$, find $\mathcal{L}\left[\displaystyle\int_0^t \int_0^t (\sin 5t)\, dt\, dt\right]$ by the following methods.

(a) Integrate twice, using the limits shown and then find the Laplace transform of that result.

(b) Use the integral theorem and compare the result with that of part (a).

1.39. If $\mathcal{L}[f(t)] = \dfrac{s^2 + 5}{s(s^2 - 6s + 9)}$, what is $\mathcal{L}[e^{-3t}f(t)]$?

Find the inverse Laplace transforms.

1.40. (a) $\dfrac{e^{-2s}}{s^2}$ **(b)** $\dfrac{8e^{-3s}}{s^2 + 4}$

1.41. (a) $\dfrac{se^{-2s}}{s^2 + 3s + 2}$ **(b)** $\dfrac{e^{-3s}}{s^2 - 2s + 5}$

1.42. (a) $\dfrac{se^{-0.8\pi s}}{s^2 + 25}$ **(b)** $\dfrac{(s + 1)e^{-\pi s}}{s^2 + s + 1}$ **(c)** $\dfrac{e^{-5s}}{(s - 2)^4}$

1.43. If $\mathcal{L}[\sinh t] = 1/(s^2 - 1)$, what is $\mathcal{L}[\sinh 2t]$?

Section 1.15

1.44. If $\mathcal{L}[f(t)] = \dfrac{2s + 1}{(s + 1)(s + 2)}$, find $\mathcal{L}[f'(t)]$.

1.45. Find $f(0+)$ and $f(\infty)$ by using the initial and final value theorems for

$$F(s) = \frac{24}{(s - 2)^5}$$

1.46. Repeat Problem 1.45 for all parts of Problem 1.29 and compare your results by substituting zero and infinity into the answers for Problem 1.29.

1.47. Repeat Problem 1.45 for the following functions.

(a) $F(s) = \dfrac{51}{s^5 + 2s^3 + 17s}$ **(b)** $F(s) = s + \dfrac{27s^4 + 5s^3 + 5s^2}{22s^5 + 2s^3}$

2

applications of laplace transforms

After completing this chapter, you should be able to:

- Write loop and nodal equations in integrodifferential form of a network, whose schematic is given.
- Transform an integrodifferential network equation into the s-domain, solve the equation for the variable in question, and transform the result back into the time domain.
- Write loop and nodal equations in the s-domain for a given RLC network with initial conditions, solve for the variable in question, and convert back into the time domain.
- Define the transfer function for a given circuit and find it.
- Find the poles and zeros for a given transfer function and determine its stability.
- Find the impulse or step response with arbitrary constants for a circuit if only the pole–zero plot is given.

- Determine the damped natural frequency and whether it is an underdamped, overdamped, or critically damped circuit for an *RLC* series or parallel circuit.
- Find the rise time, percent overshoot, and settling time by observing a graph of the time response.
- Obtain the gain equation and phase equation if $G(s)$ is given.
- If the transfer function is given, construct a Bode plot of the gain magnitude or phase by proper use of asymptotes.

INTRODUCTION

Most studies of circuit analysis consider only steady-state dc or steady-state sinusoidal ac circuits, with the ac case solved for just one frequency. If we want to analyze a circuit for any frequency or for any type of ac waveform (square wave, triangular wave, etc.), we need a more general method of solution; that is, the previous methods we have learned will simply not work. Also, it is apparent that if a circuit is suddenly connected to its source by turning on a switch, the circuit voltages and currents will be different than they would be if the source had been connected for a while. This is called the **transient effect**. It happens in mechanical systems as well as in dc and ac circuits. Just consider a dc voltage connected to a capacitor. We know that a capacitor cannot pass dc, yet with a switch in the circuit there is a current for a short time (transient) as the capacitor is charged. Besides dc and various waveshapes of ac and "switch-on" phenomena, there are the nonperiodic waveshapes to consider, such as exponentials and combinations of all these time functions, such as te^{-t} or $e^{-2t} \cos t$.

2.1 DIFFERENTIAL EQUATIONS

Consider the circuit in Figure 2.1. In ac circuit analysis, we would have assigned an impedance to the capacitor and inductor based on the C and L values and the frequency. But if we do not know the frequency of v_s or if it is not sinusoidal or

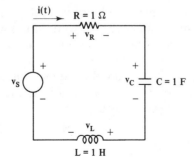

L = 1 H   **Figure 2.1**

even periodic, we cannot assign such impedances.　Instead, we must revert to the basic relationships between voltage and current for these components:

$$v_R = i_R R \tag{2.1}$$

$$v_C = \frac{1}{C} \int_{-\infty}^{t} i_c \, dt \quad \text{or} \quad i_c = \frac{dv_c}{dt} C \tag{2.2}$$

$$v_L = L \frac{di_L}{dt} \quad \text{or} \quad i_L = \frac{1}{L} \int_{-\infty}^{t} v_L \, dt \tag{2.3}$$

These relations are so basic that they will hold for every conceivable type of source: dc (switched or steady), sinusoidal (switched or steady), square wave (switched or steady), exponential, and so on.

Another law that will hold for all circuits is Kirchhoff's voltage law, KVL. Using KVL on Figure 2.1, we have

$$v_S = v_R + v_C + v_L = i_R R + \frac{1}{C} \int_{-\infty}^{t} i_C \, dt + L \frac{di_L}{dt} \tag{2.4}$$

Since this is a series circuit, $i_R = i_C = i_L = i(t)$.

We will use $i(t)$ or $v(t)$ to indicate that the current or voltage are all functions of time, in general.　Then

$$v_s(t) = i(t)R + \frac{1}{C} \int_{-\infty}^{t} i(t) \, dt + L \frac{di(t)}{dt} \tag{2.5}$$

This is called an **integrodifferential equation** since it contains an integral of $i(t)$ as well as a differential of $i(t)$.　Notice that we cannot just factor out the $i(t)$ and solve for it as we did in a purely dc steady-state case (or as we did for the steady-state ac case using phasors).　We could, however, differentiate the equation so as to get rid of the integral:

$$L \frac{d^2 i}{dt^2} + R \frac{di}{dt} + \frac{1}{C} i = \frac{dv_s}{dt} \tag{2.6}$$

This is a differential equation which can be solved for $i(t)$ if $v_s(t)$ is a known time function.　The solution can be obtained by two different methods.

The first method is the classical method, that is, solving the differential equation by calculus methods.　This can be quite tedious and requires a knowledge in this area.　It also requires that two initial conditions be given: the value of the current at $t = 0$ and the value of the derivative of the current at $t = 0$.

The second method is called **transform analysis**.　This is the easier method of the two and it is the only method that we will consider.　However, the transform method has two approaches:

1. Write the differential equations and then transform them.

2. Transform the circuit elements and then write the transform equations directly.

In the next section we show the first approach to transform analysis.

2.2 TRANSFORMING EQUATIONS

The first step in this approach to transform analysis is to write the time-domain equation. We will use the integrodifferential equation rather than the derived differential equation. For instance, for Figure 2.1, we will use Eq. 2.5 rather than Eq. 2.6. Since Eq. 2.5 contains time functions, we can take the Laplace transform of both sides of this equation:

$$\mathscr{L}[v_s(t)] = \mathscr{L}[i(t)R] + \mathscr{L}\left[\frac{1}{C}\int_{-\infty}^{t} i(t)\, dt\right] + \mathscr{L}\left[L\frac{di(t)}{dt}\right] \tag{2.7}$$

But

$$\mathscr{L}\left[\frac{1}{C}\int_{-\infty}^{t} i(t)\, dt\right] = \mathscr{L}\left[\frac{1}{C}\int_{-\infty}^{0} i(t)\, dt + \frac{1}{C}\int_{0}^{t} i(t)\, dt\right] \tag{2.8}$$

Note that the first integral on the right side of Eq. 2.8 is just the voltage on the capacitor at time $= 0$, since it is the summation (i.e., integral) of all the past accumulated charge and discharge on this capacitor. We call this the initial capacitor voltage, $v_c(0+)$. Thus

$$\mathscr{L}[v_s(t)] = R\mathscr{L}[i(t)] + \mathscr{L}[v_C(0+)] + \frac{1}{C}\mathscr{L}\left[\int_{0}^{t} i(t)\, dt\right] + L\mathscr{L}\left[\frac{di(t)}{dt}\right] \tag{2.9}$$

To find the Laplace transform of each term we must either have a knowledge of $v_s(t)$ or $i(t)$ or we must just indicate its transform. We will just indicate it as follows:

$$V_s(s) = RI(s) + \frac{v_C(0+)}{s} + \frac{1}{C}\frac{I(s)}{s} + L[sI(s) - i(0+)] \tag{2.10}$$

Notice that the second term on the right side, $v_C(0+)$, is a constant, so by T-2 it transforms to $(1/s)v_C(0+)$. The third and fourth terms involve integrals and differentials so that O-2 and O-1 from Table 1.3 are used. Notice that Eq. 2.10 is an algebraic equation in which $I(s)$ can be considered the unknown and $V_s(s)$ the known. Thus

$$I(s)\left(R + \frac{1}{sC} + sL\right) = V_s(s) - \frac{v_C(0+)}{s} + Li(0+) \tag{2.11}$$

$$I(s) = \frac{V_s(s) - [v_C(0+)/s] + Li(0+)}{R + (1/sC) + sL} \tag{2.12}$$

This is the general solution of an *RLC* series circuit in the *s* domain. Once $I(s)$ has been found, $i(t)$ can be found by taking the inverse Laplace of $I(s)$. Let us use this general result to solve a particular series circuit.

Example 2.1

Find $i(t)$ for all time after $t = 0$ for the circuit in Figure E2.1 if the initial capacitor voltage is equal to zero volts.

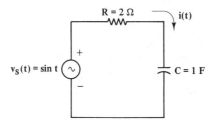

Figure E2.1

Solution   This is similar to Figure 2.1 with $L = 0$ and $v_C(0+) = 0$. Since $v_s(t) = \sin t$, $V_s(s) = 1/(s^2 + 1)$, using T-5 of Table 1.2. Therefore, from Eq. 2.12,

$$I(s) = \frac{V_s(s)}{R + 1/sC} = \frac{1/(s^2 + 1)}{2 + 1/s} = \frac{1/(s^2 + 1)}{2(s + \frac{1}{2})/s} \qquad (2.13)$$

$$= \frac{s/2}{(s^2 + 1)(s + \frac{1}{2})} = \frac{A}{s + \frac{1}{2}} + \frac{Bs + C}{s^2 + 1} \qquad (2.14)$$

By use of Eq. 1.17,

$$A = \frac{s/2}{s^2 + 1}\bigg|_{s = -\frac{1}{2}} = -\frac{1}{5}$$

To expand $(Bs + C)/(s^2 + 1)$ as a quadratic, let $s = 0$. Therefore, $C/1 = 0 - (-\frac{1}{5})/(\frac{1}{2}) = \frac{2}{5}$. Letting $s = 1$ will yield

$$\frac{\frac{1}{2}}{2(\frac{3}{2})} = \frac{-\frac{1}{5}}{\frac{3}{2}} + \frac{B + \frac{2}{5}}{2}$$

or $B = \frac{1}{5}$. Therefore,

$$\frac{s/2}{(s^2 + 1)(s + \frac{1}{2})} = -\frac{1}{5}\left(\frac{1}{s + \frac{1}{2}}\right) + \frac{\frac{1}{5}s + \frac{2}{5}}{s^2 + 1}$$

$$\mathscr{L}^{-1}\left[\frac{s/2}{(s^2 + 1)(s + \frac{1}{2})}\right] = -\frac{1}{5}e^{-t/2} + \mathscr{L}^{-1}\left[\frac{1}{5}\left(\frac{s}{s^2 + 1}\right) + \frac{2}{5}\left(\frac{1}{s^2 + 1}\right)\right] \qquad (2.15)$$

$$i(t) = -\frac{1}{5}e^{-t/2} + \frac{1}{5}\cos t + \frac{2}{5}\sin t$$

$$= -\frac{1}{5}e^{-t/2} + \frac{1}{\sqrt{5}}\sin(t + 26.5°)$$

Compare this to the result that would be achieved by ac phasor analysis. Since $\omega = 1$, V(phasor form) $= (1/\sqrt{2})\ \angle 0°$ and Z (total) $= 2 - j1 = \sqrt{5}\ \angle -26.5°$. Current, $I = V/Z$(total) $= (1/\sqrt{2})\ \angle 0° \div \sqrt{5}\ \angle -26.5° = (1/\sqrt{10})\ \angle 26.5°$. In time-domain form, this is $i(t) = (1/\sqrt{5})\sin(t + 26.5°)$. Equation 2.15 agrees with this, except that it contains an exponential function (transient response).

In Example 2.1, steady-state ac analysis would show a current of $(1/\sqrt{5})\sin (0 + 26.5°) = 0.203$ at $t = 0$, thereby producing 0.406 V across the 2-Ω resistor. Since the supply voltage actually equals zero at $t = 0$, this means that to balance the voltages in the circuit $v_C(0+)$ must be -0.406 V. But it was stated as a given fact that $v_C(0+) = 0$ at $t = 0$. This could occur if we had an uncharged capacitor and the source was connected to the circuit by throwing a switch at $t = 0$. Since a capacitor preceded by a resistor cannot charge to any voltage instantaneously, this would mean that the voltage would have to approach the ac sine wave value gradually. Since the time constant of this circuit is $RC = 2$ seconds, it would take about 10 seconds for the v_C to agree with the v_C calculated by steady-state techniques. This is exactly what the first term of the exact solution (Eq. 2.15) provides. So the first term is called the transient response and the second term is called the steady-state response. The true response is the sum of both of these.

For more complex circuits with several meshes and sources, mesh or nodal equations can be written by using Eqs. 2.1, 2.2, and 2.3. Since these equations contain not just voltages and currents, but differentials of voltages and currents, determinants cannot be used to solve these equations directly. Instead, each of these equations is first transformed into the s domain by Laplace methods. Then these s-domain equations can be solved by determinants. The final solution in the s-domain is then converted to the time domain by taking the inverse Laplace transform.

Example 2.2

Solve the following differential equation by Laplace transforms: $16x(t) + 2dx(t)/dt = 80$, if $x(0) = 4$.

Solution   The problem is to find a specific function, $x(t)$, which will satisfy this equation for all values of time. Convert this to the s-domain; we have $16X(s) + 2[sX(s) - 4] = 80/s$, using O-1 and T-2. Then $X(s)(16 + 2s) = 80/s + 8$, and

$$X(s) = \frac{(80/s) + 8}{2(s + 8)} = \frac{4s + 40}{s(s + 8)} = \frac{5}{s} + \frac{-1}{s + 8}$$

Therefore, $x(t) = 5 - e^{-8t}$.

To check this, just substitute this solution into the original equation:

$$16(5 - e^{-8t}) + 2(0 + 8e^{-8t}) = 80$$

$$80 - 16e^{-8t} + 16e^{-8t} = 80 \qquad \text{(checks)}$$

2.3 TRANSFORMING CIRCUIT ELEMENTS

As stated previously, the second method of transform analysis of circuits involves transforming the elements themselves. This does not mean that since resistance, R, is a constant that its transformed impedance is R/s. On the contrary, by transformed impedance we mean an expression for R that will give the proper relationship between $V(s)$ and $I(s)$. We have seen that from Eq. 2.10, the relation is $V(s) = RI(s)$, since R was just a multiplying constant of $i(t)$. Thus Figure 2.2a shows the time-domain circuit and Figure 2.2b shows the transformed circuit. Ohm's law applies to both circuits.

Consideration of capacitors is a bit more complicated. Capacitor voltage is given by Eq. 2.2, which can be split into two integrals, as was done in Eq. 2.8. Notice that this last integral is the summation of all the charge and discharge history of the capacitor prior to $t = 0$, so it must be just the capacitor voltage at $t = 0+$. This means that the actual capacitor (Figure 2.3a) can be modeled by a dc voltage source in series with a capacitor that is uncharged to $t = 0$ (Figure 2.3b). The transform of these two voltages is indicated in Eq. 2.8 and is given explicitly in Eq. 2.10 (second and third terms of right side). This looks as if the capacitor is a resistor whose "resistance" is $1/sC$. Figure 2.3d is the circuit with all the elements transformed. Notice that $V_s(s)$ is not divided by s because $v_s(t)$ was not necessarily a constant but is unknown at this point, so that $V_s(s)$ is an unknown transform expression also.

Figure 2.3c and its transformed circuit, Figure 2.3e, are the Norton equivalents of Figure 2.3b and d. Figure 2.3e could also be obtained by transforming the elements in Figure 2.3c. Figure 2.3d is used when writing mesh equations and Figure 2.3e is used when writing nodal equations.

In dealing with inductors, we can split up the second part of Eq. 2.3 as follows:

$$i_L(t) = \frac{1}{L} \int_{-\infty}^{0} v_L(t)\, dt + \frac{1}{L} \int_{0}^{t} v_L(t)\, dt = i_L(0+) + \frac{1}{L} \int_{0}^{t} v_L(t)\, dt \qquad (2.16)$$

This means that the actual inductor of Figure 2.4a can be modeled by a parallel combination of an inductor that is unfluxed at $t = 0$ and a dc current source, as shown in Figure 2.4b. The Thévenin equivalent of this is shown in Figure 2.4c. Transforming the elements in these two circuits yields Figure 2.4d and e. Thus L in a transformed circuit behaves as if it had a resistance of sL and the voltage across

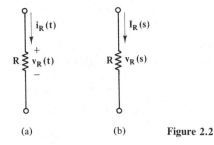

(a)   (b)   **Figure 2.2**

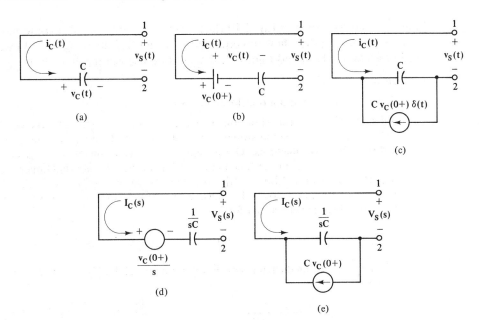

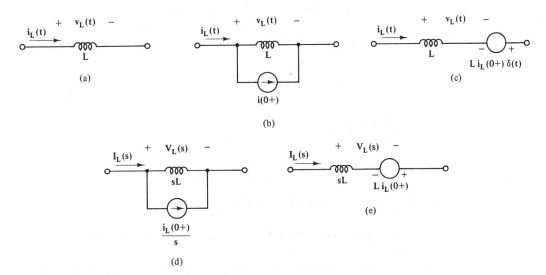

Figure 2.4

Figure 2.4c is as shown in Eq. 2.10 (last term). Figure 2.4e also could have been obtained by taking the Thévenin equivalent of Figure 2.4d. Figure 2.4d would be used in nodal analysis and Figure 2.4e would be used in mesh analysis.

Example 2.3

Solve Example 1.8 using transform analysis.

Solution   Since the capacitor voltage is zero prior to $t = 0$, the dc voltage of Figure 2.3b is missing. Since the switch is flipped at $t = 2$, the applied voltage $= (\sin \pi t/4)[u(t - 2)]$. We might use O-3 to find the Laplace of this voltage, except that O-3 requires a $(t - 2)$ in the function as well as in the unit step multiplier. But it is apparent that $(\sin \pi t/4)[u(t - 2)]$ is the same as $[\cos \pi(t - 2)/4]u(t - 2)$, which is a true shifted version of $\cos \pi t/4$. Since the transform of $\cos \pi t/4$ is $s/[s^2 + (\pi/4)^2]$, by O-3 the shifted version's transform is

$$V_s(s) = \frac{se^{-2s}}{s^2 + (\pi/4)^2}$$

Thus the transformed circuit is as shown in Figure E2.3. Using "Ohm's law" for this circuit,

$$I(s) = \frac{V_s(s)}{Z(s)} = \frac{\dfrac{se^{-2s}}{s^2 + (\pi/4)^2}}{\dfrac{1}{2s}} = \frac{2s^2e^{-2s}}{s^2 + (\pi/4)^2} = \frac{2\left[e^{-2s}\left(s^2 + \left(\dfrac{\pi}{4}\right)^2\right) - e^{-2s}\left(\dfrac{\pi}{4}\right)^2\right]}{s^2 + (\pi/4)^2}$$

$$= 2\left[(1)e^{-2s} - \frac{\pi}{4}\left(\frac{(\pi/4)e^{-2s}}{s^2 + (\pi/4)^2}\right)\right]$$

Therefore,

$$i(t) = 2\delta(t - 2) - \frac{\pi}{2}\left[\sin\frac{\pi}{4}(t - 2)\right]u(t - 2)$$

which agrees with the answer to Example 1.8.

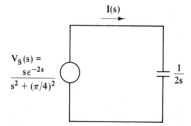

$$V_S(s) = \frac{se^{-2s}}{s^2 + (\pi/4)^2}$$

$I(s)$

$\dfrac{1}{2s}$

Figure E2.3

In summary, to solve a circuit by transform mesh analysis:

1. Change all current sources to voltage sources and transform these sources by using Table 1.2.

2. Convert capacitors and inductors to their transform equivalent by using Figures 2.3d and 2.4e, respectively.

3. Solve the circuit by writing mesh equations using the transformed circuit values from steps 1 and 2.

To solve a circuit by transform nodal analysis:

1. Change all voltage sources to current sources and transform these sources using Table 1.2.
2. Convert capacitors and inductors to their transform equivalents using Figures 2.3e and 2.4d.
3. Solve the circuit by writing nodal equations using the transformed circuit values found in steps 1 and 2. Note that nodal equations require using the reciprocal of the impedance values.

Example 2.4

Find the voltage across the capacitor in the circuit shown in Figure E2.4a for $t > 0$.

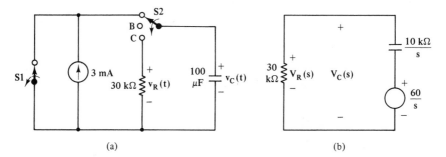

(a)   (b)

Figure E2.4

Solution   Prior to $t = -7$, S1 and S2 are as shown. At $t = -7$, S1 opens; at $t = -5$, S2 moves to B; and at $t = 0$, S2 moves to C.   Because the capacitor is shorted prior to $t = -7$, $v_c(-7) = 0$.   When S1 opens, 3 mA flows onto the capacitor for 2 s, thereby charging it with 6 mC.   Therefore, the voltage rises to 6 mC/100 μF, or 60 V.   From $t = -5$ to $t = 0$, the capacitor is not connected to anything, so that $v_c(0) = 60$ V.   After $t = 0$, the capacitor discharges through R, so that $v_c(t) = v_R(t)$. The s-domain equivalent circuit shown in Figure E2.4b is per Figure 2.3d.   Using VDR yields

$$V_R(s) = \frac{60}{s}\left(\frac{30 \text{ k}\Omega}{30 \text{ k}\Omega + 10 \text{ k}\Omega/s}\right) = \frac{60}{s + \frac{1}{3}}$$

so that

$$v_c(t) = v_R(t) = 60e^{-t/3}$$

Notice that this answer depends on the voltage on the capacitor at $t = 0$ (which is 60 V), but it does not matter in what manner this 60 V was produced.

The following example shows a complex circuit solved by mesh equations for $I_A(s)$. To find the inverse Laplace of $I_A(s)$ requires factorization of the denominator. With a denominator of this order, it is virtually impossible without a computer program. At this point, the ROOT FINDER and PART FRAC computer programs, given in Appendix B, are used.

Example 2.5

Convert the time-domain circuit shown in Figure E2.5a to an s-domain circuit so that mesh equations can be written.

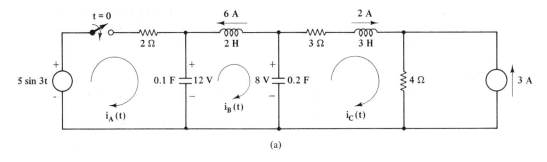

(a)

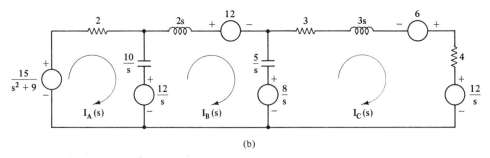

(b)

Figure E2.5

Solution The initial voltages on the capacitors and currents through the inductors must be converted to voltage sources as in Figures 2.3d and 2.4e, respectively. Other sources are converted according to Table 1.2. When this is done, the circuit shown in Figure E2.5b is the result. The mesh equations for this circuit are

$$\left(2 + \frac{10}{s}\right)I_A - \frac{10}{s}I_B - 0I_C = \frac{15}{s^2 + 9} - \frac{12}{s}$$

$$-\frac{10}{s}I_A + \left(\frac{10}{s} + 2s + \frac{5}{s}\right)I_B - \frac{5}{s}I_C = \frac{12}{s} - 12 - \frac{8}{s}$$

$$0I_A - \frac{5}{s}I_B + \left(\frac{5}{s} + 3 + 3s + 4\right)I_C = \frac{8}{s} + 6 - \frac{12}{s}$$

Solving for $I_A(s)$ by determinants:

$$I_A(s) = \cfrac{\begin{vmatrix} \dfrac{-12s^2 + 15s - 108}{s(s^2 + 9)} & -\dfrac{10}{s} & 0 \\[3mm] \dfrac{-12s + 4}{s} & \dfrac{2s^2 + 15}{s} & -\dfrac{5}{s} \\[3mm] \dfrac{6s - 4}{s} & -\dfrac{5}{s} & \dfrac{3s^2 + 7s + 5}{s} \end{vmatrix}}{\begin{vmatrix} \dfrac{2s + 10}{s} & -\dfrac{10}{s} & 0 \\[3mm] -\dfrac{10}{s} & \dfrac{2s^2 + 15}{s} & -\dfrac{5}{s} \\[3mm] 0 & -\dfrac{5}{s} & \dfrac{3s^2 + 7s + 5}{s} \end{vmatrix}}$$

Solving for $I_A(s)$ is very tedious, but when done, yields

$$I_A(s) = \frac{-6s^6 - 36.5s^5 - 151.5s^4 - 433.917s^3 - 953.75s^2 - 897.5s - 450}{s(s^2 + 9)(s^4 + 7.333s^3 + 20.833s^2 + 38.333s + 37.5)}$$

Using ROOT FINDER, the quartic in the denominator is factored:

denominator $= (s + 0.671 + j2.025)(s + 0.671 - j2.025)(s + 2.138)(s + 3.852)$

The other denominator factors are: $s(s + j3)(s - j3)$. Then, entering this information along with the numerator polynomial information into the program called PART FRAC yields the numerators for the partial fractions. Then, using T-4, we get

$$i_A(t) = -1.33 + (0.139 + j0.043)e^{-3jt} + (0.139 - j0.043)e^{3jt} + (-1.83$$

$$- j1.08)e^{-(0.671 + j2.025)t} + (-1.83 + j1.08)e^{-(0.671 - j2.025)t} + 2.438e^{-2.138t}$$

$$- 3.729e^{-3.852t}$$

Using the trigonometric identities, we finally get

$$i_A(t) = -1.33 + 2.438e^{-2.138t} - 3.729e^{-3.852t} + 0.291 \sin$$

$$(3t + 72.8°) + 4.24e^{-0.671t} \sin (2.025t - 120.5°)$$

Example 2.6

Solve the circuit shown in Figure E2.6a for $i_R(t)$ in the 2-Ω load by converting the rest of the circuit to its Thévenin equivalent first.

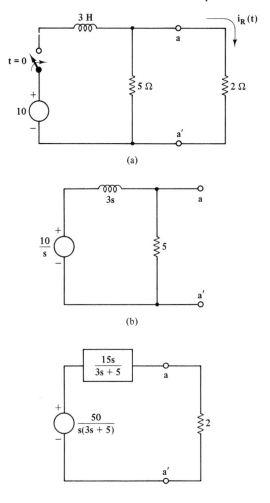

(a)

(b)

(c)   **Figure E2.6**

Solution   The portion to be thevenized is shown in its transformed state in Figure E2.6b.   The Thévenin impedance is $3s$ in parallel with 5.   This is $15s/(3s + 5)$. The Thévenin source is by the voltage-divider rule: $(10/s)[5/(3s + 5)]$.   This reduces to $50/[s(3s + 5)]$.   The final circuit is shown in Figure E2.6c.   Analyzed as a series circuit, the transform current is found:

$$I(s) = \frac{50/[s(3s + 5)]}{[15s/(3s + 5)] + 2} = \frac{50}{s(21s + 10)} = \frac{-5}{s + 10/21} + \frac{5}{s}$$

from which,

$$i(t) = 5(1 - e^{-(10/21)t}).$$

Example 2.7

In Example 2.6, we assumed there that was no current through the inductor at $t = 0$. Repeat the example, except assume that the load is connected at $t = 0$, but that the inductor was connected always (so its current had reached a steady value).

Solution   In this case we must put an additional voltage source of $6\delta(t)$ in series with the inductor before thevenizing it (see Figure E2.7a) to account for the 2-A steady current prior to $t = 0$. When converted to the s-domain, this is just 6, as shown in the circuit in Figure E2.7b. Then the new Thévenin voltage we get is

$$V_{th}(s) = \left(\frac{10}{s} + 6\right)\frac{5}{3s + 5} = \frac{10}{s}$$

When the 2-Ω load is added at $t = 0$, the current becomes

$$I(s) = \frac{10/s}{15s/(3s + 5) + 2} = \frac{10s/7 + 50/21}{s(s + 10/21)} = \frac{5}{s} + \frac{-25/7}{s + 10/21}$$

Therefore, $i(t) = 5 - \frac{25}{7}e^{-10t/21}$.

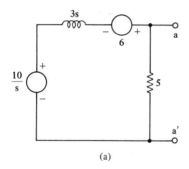

(a)

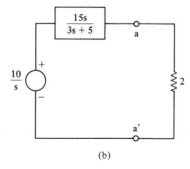

(b)   **Figure E2.7**

2.4 TRANSFER FUNCTION

A circuit is usually characterized by one or more independent sources, a number of linear passive elements, and one or more voltages or currents that can be considered as outputs. In cases where such systems can be considered as lumped, linear, time-invariant systems, the concept of a transfer function, that is, a relation between input and output, is very useful.

In cases where such a system has one input (a voltage or a current source) and one output (a voltage or current), the system can be defined by one transfer function, $G(s)$:

$$G(s) = \frac{\text{output}(s)}{\text{input}(s)} \tag{2.17}$$

In cases where the system has several inputs or several outputs, more such transfer function (ratios) exist. For instance, if a series circuit has a voltage input, $V_{in}(s)$, and a resistor and a capacitor within initial charge voltage, $v_c(0+)$ at $t = 0$, we know that the capacitor can be modeled as an uncharged capacitor in series with a dc voltage source, $v_c(0+)$, which is $v_c(0+)/s$ in the s domain. If the output is the series current, $I(s)$, then, in addition to the transfer function, $G_1(s) = I(s)/V_{in}(s)$ due to the input voltage, there is a second transfer function, $G_2(s) = I(s)/[v_c(0+)/s]$, due to the voltage source equivalent of the initially charged capacitor. The transfer function for each source and the output from each source can be found separately and added to give the total $I(s)$, using the theorem of superposition. This total $I(s)$ can be converted to the time domain to give the total time-domain current.

Consider the circuit in Figure 2.5a and its transformed form, Figure 2.5b. We will assume that there is no initial current. If $v_1(t) = 2$ V dc, combining this with the switch will yield $v_1(t) = 2u(t)$ and $V_1(s) = 2/s$. Using the voltage-divider rule, we have

$$V_o(s) = \frac{2}{s}\frac{2}{3s+4} = \frac{4}{s(3s+4)} = \frac{1}{s} + \frac{-1}{s+\frac{4}{3}}$$

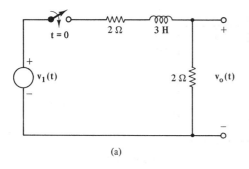

(a)

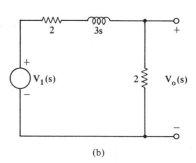

(b)

Figure 2.5

so that $v_o(t) = 1 - e^{-4t/3}$. The ratio

$$\frac{v_o(t)}{v_1(t)} = \frac{1 - e^{-4t/3}}{2}$$

and the ratio

$$\frac{V_o(s)}{V_1(s)} = \frac{4/[s(3s + 4)]}{2/s} = \frac{2}{3s + 4}$$

If, now, $v_1(t)$ is changed to 4 V dc, then $V_1(s) = 4/s$, and using the VDR, we have

$$V_o(s) = \frac{4}{s}\frac{2}{3s + 4} = \frac{2}{s} + \frac{-2}{s + \frac{4}{3}}$$

so that $v_o(t) = 2 - 2e^{-(4/3)t}$.

The ratio

$$\frac{v_o(t)}{v_1(t)} = \frac{2 - 2e^{-4t/3}}{4} = \frac{1 - e^{-4t/3}}{2}$$

and the ratio

$$\frac{V_o(s)}{V_1(s)} = \frac{8[s(3s + 4)]}{4/s} = \frac{2}{3s + 4}$$

Next, if $v_1(t)$ is changed to sin t, $V_1(s) = 1/(s^2 + 1)$, so, using VDR, we have

$$V_o(s) = \frac{1}{s^2 + 1}\left(\frac{2}{3s + 4}\right) = \frac{A}{s + \frac{4}{3}} + \frac{Bs + C}{s^2 + 1}$$

Therefore,

$$A = \frac{2}{3}\left(\frac{1}{s^2 + 1}\right)\bigg|_{s = -4/3} = \frac{2}{3}\frac{1}{\frac{16}{9} + 1} = 0.24$$

and

$$v_o(t) = 0.24e^{-4t/3} + f_2(t)$$

where $f_2(t)$ is due to the quadratic term. For the quadratic, $\alpha = 0$ and $\omega = 1$ and $Q(s) = 2/(3s + 4)$, which when evaluated at the pole $s = j$ is

$$\frac{2}{3j + 4} = 0.4\ \underline{/-37°} = M\underline{/\theta}$$

Using the "trick" formula yields

$$f_2(t) = 0.4 \sin (t - 37°)$$

so

$$v_o(t) = 0.24e^{-4t/3} + 0.4 \sin (t - 37°)$$

$$\frac{v_o(t)}{v_1(t)} = \frac{0.24e^{-4t/3} + 0.4 \sin (t - 37°)}{\sin t}$$

$$\frac{V_o(s)}{V_1(s)} = \frac{[1/(s^2 + 1)][2/(3s + 4)]}{1/(s^2 + 1)} = \frac{2}{3s + 4}$$

Notice that $v_o(t)/v_1(t)$ is the same in the first two cases but not the third. So, in general, no conclusion can be drawn. However, $V_o(s)/V_1(s)$ is the same for all three inputs. It can be shown that for any given circuit, $V_o(s)/V_1(s)$ will be the same, regardless of the input. This is a very powerful statement. It means that $V_o(s)/V_1(s)$ depends only on the circuit element values, not on the value or nature of the source. This ratio is called $G(s)$ (see Eq. 2.17) and is given the name **transfer function**.

Occasionally, we will come across complicated circuits in which none of the elements are given; the circuit is shown as merely a collection of functional blocks connected in some way. When this happens, it is just assumed that these blocks perform some operation on the voltages of the input lines and the result appears on the output of the block. A triangle might be used to indicate an amplifier with the gain shown in the block. Another block may have a $G(s)$ in it, showing the ratio of the output to the input. Another might be a circle with a plus sign to show that the input voltages are just added. Still other blocks might indicate that the input is integrated or differentiated. The following example illustrates several of these concepts.

Example 2.8

Find the transfer function, $G(s) = V_o(s)/V_1(s)$ for the functional block diagram shown in Figure E2.8. Assume that $v_1(0) = 0$.

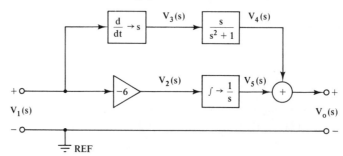

Figure E2.8

Solution With the assumption above, we see that by using O-1, the differentiator becomes s; that is, the output of that block is just s times the input, provided that everything is in the s-domain. Also from O-2, the integrator block is $1/s$. The multiplication block and the addition block will work in the s-domain also (see Eqs. 1.8 and 1.9). Therefore, the following relations result:

$$V_2(s) = -6V_1(s)$$

$$V_5(s) = \frac{1}{s} V_2(s) = \frac{1}{s} (-6V_1(s)) = \frac{-6}{s} V_1(s)$$

$$V_3(s) = sV_1(s) - v_1(0) = sV_1(s) - 0 = sV_1(s)$$

$$V_4(s) = \frac{s}{s^2 + 1} V_3(s) = \frac{s^2}{s^2 + 1} V_1(s)$$

$$V_0(s) = V_4(s) + V_5(s) = V_1(s) \left(\frac{s^2}{s^2 + 1} - \frac{6}{s} \right)$$

$$G(s) = \frac{V_0(s)}{V_1(s)} = \frac{s^3 - 6s^2 - 6}{s^3 + s}$$

2.5 IMPULSE AND STEP RESPONSE

Since the transfer function, $G(s)$, does not represent any voltage or current, but rather a ratio between two such items, the inverse of $G(s)$, which we will call $g(t)$, would at first glance seem to have no physical significance. But let us consider a very special input, the unit impulse, $\delta(t)$. Its Laplace transform is 1. Therefore, the output of the circuit (in the s-domain) is 1 times $G(s)$, or just simply $G(s)$. This means that the output in the time domain is $g(t)$. Thus $g(t)$ has the meaning: output when the input is a unit impulse, or as it is called, the **impulse response**. From a practical point of view, this means that if you study a circuit's response to an impulse, you will have enough information to find the circuit's response to any input. But it is often impractical to apply an impulse to a circuit.

A more practical input is a step function (dc input turned on at $t = 0$). The response to this type of input is called $h(t)$, the step response. If this is known, the step response in the s-domain, $H(s)$, can easily be calculated. Then since the Laplace of a step input is $1/s$, $G(s)$ can be calculated as follows:

$$G(s) = \frac{H(s)}{1/s} = sH(s) = [sH(s) - h(0+)] + h(0+) \qquad (2.18)$$

Taking the inverse of both sides and using theorem O-1 yields

$$g(t) = h'(t) + h(0+)\delta(t) \qquad (2.19)$$

or

$$h(t) = \int_0^t g(t)\, dt. \qquad (2.20)$$

2.6 DRIVING-POINT FUNCTIONS

If the output and the input are at the same point in the circuit, there are only two possibilities for the $G(s)$. Either $G(s) = V(s)/I(s)$, in which case $G(s)$ is called the **driving-point impedance**, $Z(s)$; or, $G(s) = I(s)/V(s)$, in which case $G(s)$ is called the **driving-point admittance**, $Y(s)$. Thus if the circuit is driven by a voltage source at the input, the input current could be considered the output. This may seem a bit odd, but if you consider any cause to be an input and any effect to be an output, the concept should be more acceptable.

So far, we have used the transfer function concept to describe a circuit that is already in existence; that is, we are analyzing a given circuit. It should also be possible to design a circuit if the transfer function is given. This is called **realizing** the circuit. Another name is the **art of synthesis**. Nowadays, of course, it is more a science than an art. In fact, it is so complex that entire books are devoted to this one subject. Since the transfer function could be a driving-point impedance, a transfer admittance, a voltage transfer ratio, or any other of the various combinations of output and input, there are many sides of this subject to consider. A first consideration is to determine theoretically if the given transfer function can be realized. It can be proven that some functions cannot be realized at all.

For instance, in the circuit of Figure 2.6a, $Z_{dp}(s) = 3s + 4 + 1/2s = (6s^2 + 8s + 1)/2s$; so it is obvious that if $Z_{dp}(s) =$ a polynomial divided by a single term, the realization is a series circuit. Similarly, in Figure 2.6b,

$$Z_{dp}(s) = \frac{1}{1/3s + 1/4 + 2s}$$

$$= \frac{12s}{24s^2 + 3s + 4}$$

In this case, it is obvious that if the numerator is a single term and the denominator is a polynomial in s, the realization is a parallel circuit.

If the degree of the single term differs from that of the polynomial by more than one, the circuit cannot be realized, since this would require elements to have

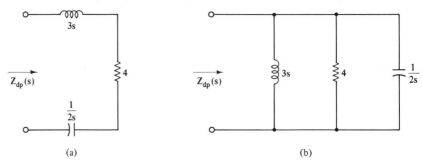

(a)   (b)

Figure 2.6

impedances such as s^2 or $1/s^2$, which do not exist (we have just R, $1/sC$, and sL). Also, if any of the signs are different, no circuit can be realized, since we can have no negative impedance. So we see that several restrictions apply, which will tell us if these series or parallel circuits are solvable.

Example 2.9

Realize $Z_{dp}(s) = 3s/(s^2 + 2s + 2)$.

Solution Since this is recognized as a parallel circuit, we must force the numerator to be one, by dividing by $3s$:

$$Z_{dp}(s) = \frac{1}{s/3 + 2/3 + 2/3s}$$

or

$$Y_{dp}(s) = \frac{s}{3} + \frac{2}{3} + \frac{2}{3s}$$

as shown in Figure E2.9. This is the sum of three admittances, from which it is deduced that

$$\frac{1}{R} = \frac{2}{3} \quad \text{or} \quad R = \frac{3}{2}\Omega$$

$$sC = \frac{s}{3} \quad \text{or} \quad C = \frac{1}{3}F$$

$$\frac{1}{sL} = \frac{2}{3s} \quad \text{or} \quad L = \frac{3}{2}H$$

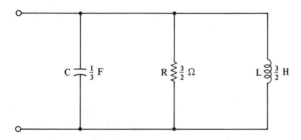

$$C = \frac{1}{3}F \qquad R = \frac{3}{2}\Omega \qquad L = \frac{3}{2}H$$

Figure E2.9

When we turn our attention to combination circuits, we will get polynomials in both numerator and denominator. In that case, even more restrictions apply as far as the realizability of the circuit is concerned.

If the person giving the problem guarantees that a solution exists, one method that sometimes works is to divide the numerator by the denominator (if the degree of the numerator is greater or equal to the degree of the denominator), or divide the denominator by the numerator (if the opposite is true). This division will give a single term plus a remainder polynomial fraction, which can be inverted and the process repeated, thereby yielding what is called a **continuing fraction**. This process is best illustrated by an example.

Example 2.10

Realize

$$Z_{dp}(s) = \frac{12s^2 + 1}{24s^3 + 6s} = \frac{1}{\dfrac{24s^3 + 6s}{12s^2 + 1}}$$

$$= \frac{1}{2s + \dfrac{4s}{12s^2 + 1}} = \frac{1}{2s + \dfrac{1}{3s + 1/4s}}$$

Solution   This is recognized as a ladder, as shown in Figure E2.10. The total impedance of any ladder network can always be expressed in this "continued" fraction form.

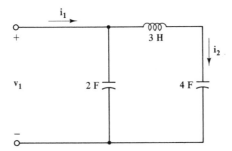

Figure E2.10

However, just because this procedure worked in this example does not mean
that it will work in another example. For the general case, a different procedure
is used which is guaranteed to work in all cases if a solution does indeed exist.
However, it is quite complex. The procedure would require several chapters to
explain, so it will not be covered in this book.

2.7 POLES AND ZEROS

Since a transfer function is, in general, a ratio of two polynomials in s, the information contained in the transfer function can be represented in a different manner.
For instance, if we were to tabulate a list of the roots of the numerator and the
roots of the denominator, this would be almost enough to reconstruct the transfer
function. If the transfer function was

$$G(s) = \frac{6s(s + 2)(s + 5)}{(s + 1)(s + 4)}$$

the roots of the numerator polynomial would be 0, -2, -5. If any of these values
are substituted in the equation above, the numerator would become zero, causing
$G(s)$ to become zero also. Because $G(s)$ becomes zero under these conditions,
these values are called the "zeros" of $G(s)$.

The roots of the denominator are -1 and -4. These are the values, if substituted for s, that would make the denominator become zero, and hence $G(s)$ would become infinite. Thus these values could be called the **zeros** of the denominator polynomial, or the **poles** of $G(s)$, since as s approaches these values, the value of $G(s)$ rises sharply toward infinity. This would look something like a pole if $G(s)$ were plotted versus s.

Thus the poles and zeros of the transfer function, $G(s)$, are tabulated as follows:

Poles	Zeros
-1	0
-4	-2
	-5

If we did not know $G(s)$, but were given the table, we now should be able to reconstruct $G(s)$. The numerator would consist of factors of the form $(s - z)$, where z is a zero from the table. Similarly, the denominator would consist of factors such as $(s - p)$, where p is a pole from the table. By this procedure, the entire $G(s)$ could be found, except for the multiplying constant [6 in the previous $G(s)$]. If the person giving the problem had told us that $G(2) = \frac{56}{3}$ as an extra piece of information, the multiplying constant, k, could be found as follows:

$$k \frac{s(s + 2)(s + 5)}{(s + 1)(s + 4)} = \frac{56}{3}$$

When $s = 2$,

$$k \frac{2(2 + 2)(2 + 5)}{(2 + 1)(2 + 4)} = \frac{56}{3}$$

Therefore, $k = 6$.

So this section has given an alternative way of expressing the same information that is contained in the transfer function itself. Of course, it is realized that not all the poles and zeros will be real numbers; some may be imaginary or even complex numbers. Since all real, imaginary, and complex numbers can be diagrammed on the complex plane, this offers yet another alternative to the representation of this information. This is called a **pole–zero plot**. These points on the complex plane will be indicated by a cross if they are a pole, and by a small circle if they are a zero. In the case of two or more zeros (multiple zeros), two or more concentric circles can be used (some texts use a little 2 next to the circle to show a double zero). In the case of multiple poles, a tilted pound sign (#) could be used for a double pole (or a small 2 next to the x, as in some books).

So, to represent a function on a pole–zero plot would require the following steps:

1. Factor the numerator and denominator into linear factors [this means that $(s^2 + 4)$ becomes $(s + j2) (s - j2)$].
2. Cancel any common factors between numerator and denominator.
3. Set each factor equal to zero and solve for s. This is the value that is to be plotted [a circle if it came from the numerator or a cross ($\times$) if it came from the denominator].
4. There is no way of plotting the multiplying constant.
5. In addition to the poles and zeros found by this method, there may be poles or zeros at infinity. If there are more zeros than poles, just throw in some poles at infinity to make the number of poles and zeros the same. Do the opposite if the number of poles exceeds the number of zeros.

In regard to step 5, these poles or zeros at infinity will never show up in $G(s)$ as a factor such as $(s - \infty)$, yet it can be shown that they do indeed exist; see the following example.

Example 2.11

Plot $G(s) = \dfrac{s(s^2 + 4)(s^2 - s - 6)}{7(s^2 + 2s + 2)(s + 2)}$.

Solution   This looks like it might contain five zeros and three poles; however, upon factoring, it is seen that two factors cancel:

$$G(s) = \frac{s(s + j2)(s - j2)(s - 3)(s + 2)}{7(s + 1 + j)(s + 1 - j)(s + 2)}$$

Since $(s + 2)$ appears in the numerator and in the denominator, it cancels. It would be incorrect to plot a zero and a pole at the same point. After plotting the rest of the points, the plot looks as shown in Figure E2.11. We also require two poles at infinity to get an equal number of poles and zeros. To show that they exist, substituting "infinity" for s in $G(s)$ will yield ∞/∞. To evaluate this, we must use l'Hôpital's rule,

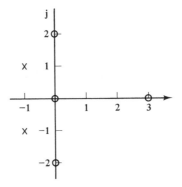

Figure E2.11

or divide numerator and denominator by a sufficient number of s factors prior to substitution. In this example, we first divide by s^2.

$$G(s) = \frac{1(1 + j2/s)(s - j2)(s - 3)}{7[1 + (1 + j)/s][1 + (1 - j)/s]}$$

When $s = \infty$ is substituted, we get

$$G(\infty) = \frac{1(1)(\infty)(\infty)}{7(1)(1)} = \frac{\infty}{7} = \infty$$

A proof that two poles at infinity are required will not be demonstrated here. We will just accept as true, step 5 in the procedure above.

Even though some of the factors (hence poles and zeros) are complex, when all these factors are multiplied together, it is seen that the coefficients are all real numbers. It is a mathematical fact that if the coefficients of the expanded polynomials are real numbers, any complex (or imaginary) poles or zeros must occur in conjugate pairs; that is, any pole or zero in the top half-plane must have a mate (a mirror image) in the bottom half-plane. In Example 2.11 the mirror image of $-1 + j$ is $-1 - j$.

2.8 TIME-DOMAIN BEHAVIOR

To find the time-domain expression when the s-domain expression is given, we saw that we usually had to expand the s-domain expression by partial fractions, thus eventually obtaining a sum of various time expressions. Examination of Table 1.2 shows that these time expressions depend on the denominator factors in the s-domain expression. For instance, if the denominator is $s^2 + 4$, we know that the result will be sinusoidal (sin or cos). This means that when looking at a pole–zero plot, it is the poles that tell us which time expressions make up the complete answer. The zeros [from the numerator of $F(s)$] affect only the magnitude of the various time expressions. Consider the following example.

Example 2.12

Find the s-domain and time-domain expressions for the $F(s)$ represented by the three pole–zero plots shown in Figure E2.12 and compare them.

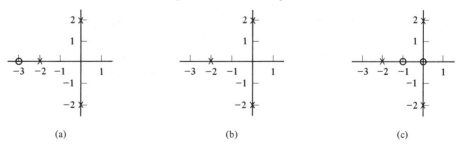

(a)   (b)   (c)

Figure E2.12

Solution   From the plots, we find the following s-domain expressions:

$$F_a(s) = k_a \frac{s + 3}{(s + 2)(s^2 + 4)}$$

$$F_b(s) = k_b \frac{1}{(s + 2)(s^2 + 4)}$$

$$F_c(s) = k_c \frac{s(s + 1)}{(s + 2)(s^2 + 4)}$$

Using partial fraction expansion and referring to Table 1.2, we obtain the following time-domain answers:

$$f_a(t) = k_a \left(\frac{1}{8}e^{-2t} - \frac{1}{8}\cos 2t + \frac{5}{8}\sin 2t \right)$$

$$f_b(t) = k_b \left(\frac{1}{8}e^{-2t} - \frac{1}{8}\cos 2t + \frac{1}{8}\sin 2t \right)$$

$$f_c(t) = k_c \left(\frac{1}{4}e^{-2t} + \frac{3}{4}\cos 2t - \frac{1}{4}\sin 2t \right)$$

A comparison shows that all the results look identical, except for the relative "weighting" of the individual terms.

Although the numerator factors [zeros of $F(s)$] account for the weighting, there is no way of telling what this is by just knowing what these zeros are; we must actually perform the partial fraction analysis to find this weighting.   However, to find the terms without the weighting is rather easy:

 1. A negative real pole (left half-plane, or LHP) yields a negative exponential, that is, a decaying term.
 2. A positive real pole (RHP) yields a positive exponential term, which will grow without bound as time passes.
 3. An imaginary pole (and its conjugate) yields a sine and/or a cosine term (or equivalently, a sine term with phase angle).
 4. A pole at the origin yields a constant.
 5. A complex pole (and its conjugate), if in the LHP, yields a damped sinusoid, that is, one that dies out gradually.
 6. n-order multiple poles yield, in addition to the response characterized by a simple pole, a term similar to the simple-pole term multiplied by $t^{n-1}/n!$ (see Examples 1.19 and 1.23).

Example 2.13

A circuit whose transfer function, $G(s)$, has the pole–zero plot shown in Figure E2.13, is excited by $v_1(t) = \sin 5t$, find its output, $v_2(t)$.

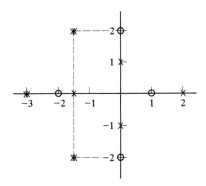

Figure E2.13

Solution   Because we know that $v_2(s) = G(s)v_1(s)$, the denominator of $v_2(s)$ must equal the denominator of $G(s)$ times the denominator of $v_1(s)$. This means that the poles of $v_2(s)$ are the poles of $G(s)$ and $v_1(s)$. The result is that $v_2(t)$ will have terms corresponding to $v_1(t)$ in addition to those obtained from the pole–zero plot of $G(s)$. Bringing all these concepts together will give the $v_2(t)$.

$$v_2(t) = k_1 e^{-1.5t} + k_2 e^{2t} + k_3 \sin(t + \theta_3) + k_4 e^{-3t} + k_5 t e^{-3t}$$

$$+ k_6 e^{-1.5t} \sin(2t + \theta_6) + k_7 t e^{-1.5t} \sin(2t + \theta_7) + k_8 \sin(5t + \theta_8)$$

In the result above, all the k and θ are constants, which can only be found by a complete analysis involving partial fractions.

2.9 STABILITY

In Example 2.13, if the input had been an impulse, then $V_1(s)$ would be $= 1$. Therefore, only $G(s)$ would have contributed denominator factors, and the resulting $v_2(t)$ would be similar to just the first seven terms of the answer given in that example. This would, of course, be called the **impulse response**.

A circuit is said to be **inherently stable**, **unstable**, or **marginally stable** if its impulse response approaches zero, infinity, or a constant nonzero value (or constant-amplitude oscillation), respectively, after a sufficiently long time. Examples are shown in Figure 2.7.

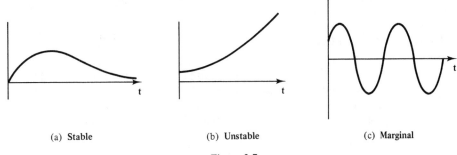

(a) **Stable**   (b) **Unstable**   (c) **Marginal**

Figure 2.7

As discussed above, the impulse response would be the first seven terms (but probably with different k and θ constants). In any case, it is seen that all the terms decrease toward zero, except for the third term (which is a constant-amplitude oscillation) and the second term (which approaches infinity as time increases). So, is this system stable, marginally stable, or unstable? The answer is that it is unstable, since it is the sum of all three, and therefore the character of $v_2(t)$ is determined by the "worst" of the seven terms. However, since the character of each of the terms is immediately obvious just by looking at the pole–zero plot, we can establish an easy procedure for determining the stability of a system:

1. If there are any poles in RHP, or if there are multiple poles on the j-axis, the system is unstable, so don't do steps 2 and 3.
2. If there are any simple poles on the j-axis, the system is marginally stable, so don't do step 3.
3. If neither of the above is true, the system is unconditionally stable.

As a practical matter, a system with just passive components will always be stable. A stable circuit is usually to be desired. A circuit with amplifiers and positive feedback can be unstable with certain values of components. This will cause undesirable oscillation. This oscillation will increase until the system becomes saturated, causing the waveform to be clipped. If designed properly, this will not destroy the circuit. In fact, these oscillations may be quite useful if the intent is to design an oscillator.

In all this discussion of stability, it is again stressed that the zeros play no role in determining stability.

Although there are normally no restrictions on the locations of the zeros of the transfer function, the situation is different for a driving-point function. If this function is a driving-point impedance consisting of just passive components, it will not be unstable. It will have no poles in the RHP. Now consider the driving-point admittance of the same circuit. It, too, must have no poles in the RHP. But since it is the reciprocal of the driving-point impedance, the poles of one are the zeros of the other. In other words, driving-point impedances of passive networks will never have zeros in the RHP. Driving-point functions of passive circuits, however, may be marginally stable, that is, contain poles or zeros of simple order on the j axis, if no resistors are involved (no damping).

2.10 NATURAL RESPONSE

In Section 2.5 we considered the response to an impulse and a step. These responses will be considered the natural response of the system. In a circuit with no electronic amplifiers or "negative" resistors, this response will contain exponentials and will die out with time. This means that all the poles will lie in the left half-plane and the system is stable. Thus the natural response will also be the transient response. If the natural response does contain sinusoids, these will be in the form of damped sinusoids, and the frequency of the damped sinusoids must be due to the components in the circuit, since there is nothing in the step or impulse input that contains any sinusoidal information. This frequency is called the **natural frequency**, ω_d.

Example 2.14

Find the impulse and step response for the circuit shown in Figure E2.14a.

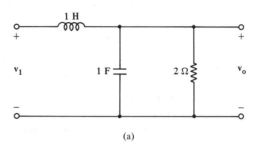

(a)

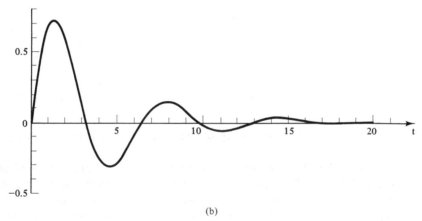

(b)

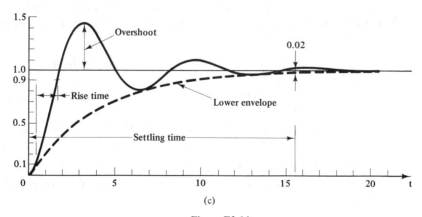

(c)

Figure E2.14

Solution   The parallel combination of resistor and capacitor is $1/(s + \frac{1}{2})$. Using VDR gives us

$$G(s) = \frac{V_o(s)}{V_1(s)} = \frac{1/(s + \frac{1}{2})}{1/(s + \frac{1}{2}) + s} = \frac{1}{s^2 + s/2 + 1}$$

Therefore, the impulse response is $g(t)$, using T-7:

$$g(t) = \frac{4}{\sqrt{15}} e^{-t/4} \sin \frac{\sqrt{15}}{4} t$$

The step response, $h(t)$, can be found from Eq. 2.20 or by Eq. 2.17:

$$h(t) = \mathcal{L}^{-1} [V_o(s)] = \mathcal{L}^{-1} \left[\frac{G(s)}{s} \right] = \mathcal{L}^{-1} \left[\frac{1}{s(s^2 + s/2 + 1)} \right]$$

This can be broken into two fractions. The first is $1/s$, whose inverse Laplace is $u(t)$. The second has a quadratic denominator and its inverse will be found using the "trick" formula $(Q(s) = 1/s)$. Since the denominator has a pole at $-\frac{1}{4} + j(\sqrt{15}/4)$, we see that

$$\alpha = \frac{1}{4} \quad \text{and} \quad \omega = \frac{\sqrt{15}}{4} \quad \text{and} \quad M\angle\theta = \left. \frac{1}{s} \right|_{s = -(1/4) + j(\sqrt{15}/4)} = 1\angle -104.5°$$

Therefore,

$$h(t) = u(t) + \frac{4}{\sqrt{15}} e^{-t/4} \sin \left(\frac{\sqrt{15}}{4} t - 104.5° \right)$$

Plots of the impulse and step response are shown in Figure E2.14b and c.

Note that the impulse input of Example 2.14 produces a rounded pulse as an output plus some "ringing." The step input produces an output that looks roughly like a step, but the rise time is gradual, there is some overshoot, and there is some ringing afterward. This is an underdamped circuit. In some circuits no ringing is produced. The subject of damping is discussed in more detail in Section 2.12. The fact that the step response is not a pure step leads to terminology used to describe this lack of agreement:

Rise time: the time required for the output to go from 10% to 90% of the final value

Settling time: the time required to get to where the magnitude stays within 2% of the final value

Overshoot: the difference in the magnitude of the peak value and the final value, expressed as a percent of the final value.

Thus, ringing is described quantitatively by quoting the settling time and the overshoot. The three values above can be measured from a plot of the step response.

Example 2.15

Find the rise time, overshoot, and settling time for Example 2.14.

Solution   Since Figure E2.14c is a plot for the step response of Example 2.14, it is seen that the rise time = $t_{90} - t_{10}$ = 1.7 − 0.4 = 1.3 seconds. Similarly, the settling time is 15 seconds and the overshoot is 13%.

2.11 *RLC* CIRCUIT RESPONSE

Let us now consider a series *RLC* circuit in which the input is 1 V dc switched on at time = 0, and the output is the resistor voltage as shown in Figure 2.8. Using VDR, we find that

$$V_o(s) = E(s) \frac{R}{sL + 1/sC + R} = \frac{R/L}{s^2 + sR/L + 1/LC} \qquad (2.21)$$

The denominator is second order but may have simple real, multiple real, or complex roots, depending on the circuit values. We consider all the possibilities. Using the quadratic formula to find the poles yields

$$s = \frac{-R/L \pm \sqrt{(R/L)^2 - 4/LC}}{2} = \frac{-R}{2L} \pm \sqrt{\left(\frac{R}{2L}\right)^2 - \frac{1}{LC}} \qquad (2.22)$$

1. *Overdamped* ($R/2L > 1/\sqrt{LC}$). In this case, the radical (discriminant) is positive, leading to two separate real roots:

$$V_o(s) = \frac{A_1}{s + \alpha_1} + \frac{A_2}{s + \alpha_2}$$
$$v_o(t) = A_1 e^{-\alpha_1 t} + A_2 e^{-\alpha_2 t} \qquad (2.23)$$

It can be shown that $A_1 = -A_2$.

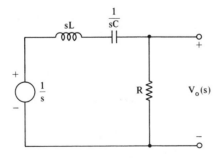

Figure 2.8

2. *Critically damped* ($R/2L = 1/\sqrt{LC}$). In this case, the discriminant is zero, leading to a double real root of the form T-10:

$$v_o(t) = Bte^{-\alpha t} \tag{2.24}$$

3. *Underdamped* ($R/2L < 1/\sqrt{LC}$). In this case, the discriminant is negative, leading to complex roots of form T-7:

$$v_o(t) = Ce^{-\alpha t} \sin \omega t \tag{2.25}$$

Since the denominator of Eq. 2.21 is of the form $s^2 + 2\alpha s + \omega_0^2$, which can be put in the form $(s + \alpha)^2 + \omega_d^2$ by completing the square, these constants can be defined as follows:

$$\alpha = \frac{R}{2L} = \text{damping constant}$$

$$\omega_0 = \frac{1}{\sqrt{LC}} = \text{undamped natural frequency}$$

$$\omega_d = \sqrt{\omega_0^2 - \alpha^2} = \text{(damped) natural frequency}$$

$$\zeta = \frac{\alpha}{\omega_0} = \text{damping ratio}$$

Thus the circuit is overdamped, critically damped, or underdamped, depending on whether α is greater than, equal to, or less than ω_0. These responses (Eqs. 2.23 to 2.25) are plotted in Figure 2.9.

Example 2.16

In Figure 2.8, let $L = 0.01$ H, $C = 0.01$ μF, and $R = 1600$ Ω. Is the response overdamped, underdamped, or critically damped, and what is its natural frequency? What must L be changed to so that the circuit will be critically damped?

Solution

$$\alpha = \frac{R}{2L} = \frac{1600}{2(0.01)} = 80,000$$

$$\omega_0 = \frac{1}{\sqrt{LC}} = \frac{1}{\sqrt{(0.01)(0.01 \ \mu F)}} = 100,000$$

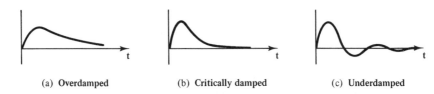

(a) Overdamped   (b) Critically damped   (c) Underdamped

Figure 2.9

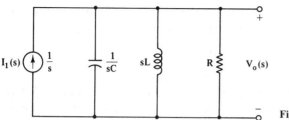

Figure 2.10

Therefore, $\alpha < \omega_0$, and this means that it is underdamped. The natural frequency is $\sqrt{100{,}000^2 - 80{,}000^2} = 60{,}000$ rad/s, or 9550 Hz. To achieve critical damping, such as in Figure 2.9b, $\alpha = \omega_0$, or $R/2L = 1/\sqrt{LC}$, or $L = R^2C/4 = 0.0064$ H.

If a parallel RLC circuit driven by a unit step current is considered, as in Figure 2.10, the voltage $V_o(s)$ is given by

$$V_o(s) = \frac{1/C}{s^2 + s/RC + 1/LC} \quad \text{and} \quad \alpha = \frac{1}{2RC} \quad (2.26)$$

ω_d and ω_o are defined the same as before. Using these values and the new definition for α, and comparing α and ω_o, it can be determined if this parallel circuit is overdamped, underdamped, or critically damped.

2.12 FORCED RESPONSE

On the other hand, if the input is a steady sinusoidal input, the output will be a steady sinusoidal response plus the transient response mentioned in Section 2.10. Thus, after sufficient time has passed, the output is just pure sinusoidal and has the same frequency as the input. This is the basis of a course on ac circuit analysis. To show that this is true, consider a circuit with a transfer function, $G(s)$. $G(s)$ can have no denominator factors, such as $(s^2 + \omega^2)$, unless the circuit is an oscillator, that is, has amplification with positive feedback. Therefore, no steady-state sinusoids will appear in the output unless contained in the input. The output $V_o(s)$ $= V_1(s)G(s)$. For instance, if the input is a sine wave of unit magnitude, then

$$V_o(s) = \frac{\omega}{s^2 + \omega^2} \, G(s) = \frac{As + B}{s^2 + \omega^2} + \left(\begin{array}{l} \text{fractions containing} \\ \text{denominator factors} \\ \text{from } G(s) \end{array} \right) \quad (2.27)$$

If we are interested only in the steady-state output, then we only want the inverse of the first part of the right side of Eq. 2.27. This can be found by using Eq. 1.23, the trick formula. From Eq. 1.22, $Q(s) = \omega G(s)$ and $M\angle\theta = \omega G(s)|_{s=j\omega} = \omega G(j\omega)$, which is really quite a simple result. Thus $M/\omega = |\omega G(j\omega)/\omega|$, or $A(\omega)$, as it is sometimes called; and $\theta = \underline{/G(j\omega)}$, or $\alpha(\omega)$, as it is sometimes called. Substituting these values into the "trick" formula yields

$$f(t) = A(\omega) \sin [\omega t + \alpha(\omega)]$$

This means that $A(\omega)$ is the steady-state sinusoidal gain of the circuit and $\alpha(\omega)$ is the phase shift of the circuit under steady-state conditions. Since both of these output quantities depend on the frequency of the input source that is causing or forcing this response, it is called the **forced response**, or **frequency response**, of the system. These quantities, when plotted on a graph, are called the **frequency plots**.

To reiterate, the quantities to be plotted, $A(\omega)$ and $\alpha(\omega)$, are first calculated as follows before plotting:

$$A(\omega) = |G(j\omega)| \quad \text{and} \quad \alpha(\omega) = \tan^{-1} \frac{\text{Im } (G(j\omega))}{\text{Re } (G(j\omega))} \qquad (2.28)$$

2.13 BODE PLOTS

There are many ways to arrange the scales for the above-mentioned frequency plots, but one of the best ways is to use a logarithmic spacing for the frequency axis and plot degrees linearly for the phase scale, but plot magnitude logarithmically or convert magnitude to decibels and plot that linearly. When this is done, the plots are called **Bode frequency plots**. By spacing the frequency logarithmically, zero frequency (dc) is infinitely far to the left, since log (0) is negative infinity; so zero frequency cannot be plotted. Also, zero magnitude cannot be plotted for the same reason; if you try to convert zero magnitude to decibels, you get negative infinity again. Yet this Bode system for plotting is quite useful.

Consider a simple three-element circuit shown in Figure 2.11. Using VDR yields

$$G(s) = \frac{V_o(s)}{V_1(s)} = \frac{100}{100 + 0.02s} = \frac{5000}{s + 5000}$$

$$G(j\omega) = \frac{5000}{5000 + j\omega} = \frac{5000}{\sqrt{25 \times 10^6 + \omega^2} \; \underline{/\tan^{-1} \omega/5000}}$$

$$= \frac{5000}{\sqrt{25 \times 10^6 + \omega^2}} \; \underline{/- \tan^{-1} \omega/5000}$$

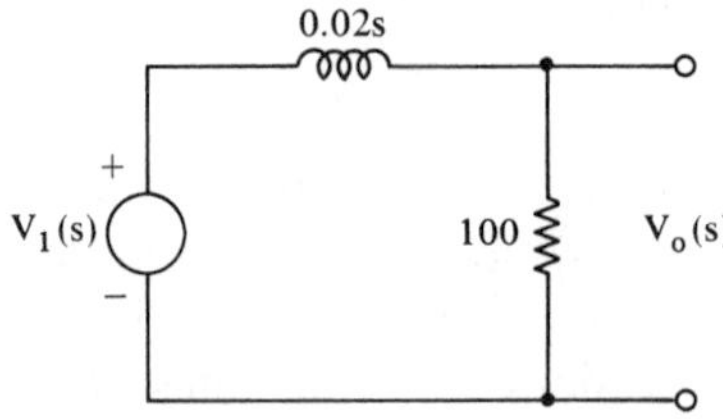

Figure 2.11

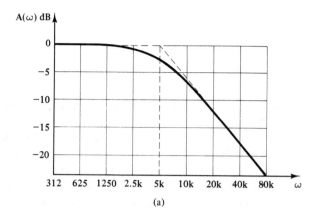

(a)

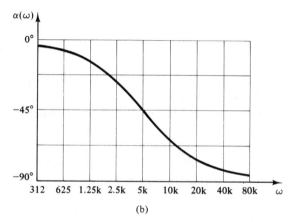

(b)

Figure 2.12

Then using Eq. 2.28, we have

$$A(\omega) = \frac{5000}{\sqrt{25 \times 10^6 + \omega^2}} \quad \text{and} \quad \alpha(\omega) = -\tan^{-1}\left(\frac{\omega}{5000}\right) \qquad (2.29)$$

The gain and phase functions can be plotted as Bode plots, as in Figure 2.12. These curves will have certain characteristics. For instance, the phase plot will be symmetrical about the $-45°$ point of this curve. The gain dB plot will consist of almost perfectly straight lines except for the blending curve in the neighborhood of $\omega_0 = 5000$ rad/s.

This means that a quick sketch can be made of the gain plot by drawing two lines that cross at $\omega_0 = 5000$ rad/s and then blending them. These lines are called **Bode asymptotes**. But why does the response approach straight lines (within 0.5 dB) if we are away from the blend point (where the asymptotes cross) by a frequency factor of 3 or $\frac{1}{3}$, say? Let's use Eq. 2.29 as an example. We see that for ω very small, $A(\omega) = 1$ and does not change much until ω approaches 5000. At $\omega = 5000$,

$$A(\omega) = \frac{5000}{\sqrt{25 \times 10^6 + (5000)^2}} = 0.707$$

which is -3 dB. This is a large error and is typical of the error at these blend points, or **breakpoints**, as they are called. However, if we go to one-third of this frequency, we will see that the gain is only about -0.5 dB in error. Now let's move to the other side of the breakpoint. At high frequency we see that $\omega \gg 5000$, so that the gain drops with frequency and is approximately $5000/\omega$. A plot of $5000/\omega$ versus ω would normally be a hyperbola, but by converting $5000/\omega$ to decibels and plotting against a ω scale with log spacing, this actually plots as a slant line and is the second asymptote shown on Figure 2.12. As long as $\omega > 3\omega_0$, the actual curve will be almost identical to this asymptote, the agreement being better as we get farther from ω_0.

The rules for plotting $A(\omega)$ using asymptotes, if $G(s)$ is given, are as follows:

1. Factor the numerator and denominator of $G(s)$ into linear factors.
2. Make a list of the magnitude of the poles and zeros of $G(s)$ in ascending order of magnitude. If some poles or zeros are complex, convert to polar and list the magnitude of these.
3. Evaluate $G(s)$ at $s \ll$ the smallest nonzero pole or zero, by substituting an arbitrarily low value of s into $G(s)$ and ignore small values that are added to larger values [thus $(s + 5)_{s=1} \approx 5$]. Convert this to decibels and plot this point.
4. If there are no poles or zeros at $s = 0$, draw a horizontal line from the plotted point to the first pole or zero. If there are poles or zeros at $s = 0$, draw a line from the plotted point at a slope of $20n$ dB/decade, where n is the number of zeros at $s = 0$, or at a slope of $-20n$ dB/decade, where n is the number of poles at $s = 0$.

TABLE 2.1 RULES FOR CHANGE IN BODE SLOPE

Poles or zeros encountered	Slope	
	dB/ decade	dB/octave
1 zero	20	6
2 zeros	40	12
3 zeros	60	18
1 pole	-20	-6
2 poles	-40	-12
3 poles	-60	-18

5. At the first breakpoint (nonzero "pole" or "zero" frequency), continue the line (see Table 2.1). Do not use the slope in the table. Instead, change the slope you had in the previous step by the amount from the table.

6. Repeat the process above as you move to the right on your plot and encounter more break frequencies (poles or zeros).

7. When you are done, you should have a continuous line containing one or more changes of direction. These are called **asymptotes**.

8. If you wish, you may now sketch in the actual response curve by following these asymptotes, except near the break frequencies, where you will make a smooth blend by staying $3n$ decibels above (or below) the breakpoints (if the pole or zero is real), where n is the multiplicity of the pole or zero in question. If the breakpoints are close together, say less than a $3:1$ ratio, you will have to stay more or less than $3n$ decibels away.

The rules for plotting $\alpha(\omega)$ using asymptotes, if $G(s)$ is given, are:

1. Same as rule 1 for $A(\omega)$.

2. Same as rule 2 for $A(\omega)$.

3. Since $\alpha(\omega) = \underline{/G(j\omega)} = $ the sum of the phase angles of all the numerator factors minus the sum of all the phase angles of all the denominator factors, it is best to plot the angle of each numerator factor versus ω, and the negative of the angle of each denominator factor as separate curves, but on the same graph. The asymptotes are horizontal lines at $0°$ and $\pm 90°$, and a slant line with a slope of about $\pm 66°$/decade at the intersection of the break frequency and $\pm 45°$. Complex poles and zeros should be handled by actually plotting a sufficient number of points near the break frequency.

4. Plot a horizontal line at $+90°$ for zeros at $s = 0$, and at $-90°$ for poles at $s = 0$.

5. Sketch the final phase plot, $\alpha(\omega)$, by adding all the phase plots above. The actual curve is blended wherever asymptotes cross each other.

Let us apply these techniques to a second-order circuit, that is, a circuit with two breakpoints.

Example 2.17

Consider the transformed circuit shown in Figure E2.17a. Make Bode plots of $A(\omega)$ and $\alpha(\omega)$ if $G(s) = V_o(s)/V_1(s)$.

Solution   By VDR,

$$G(s) = \frac{11}{s + 11 + 10/s} = \frac{11s}{s^2 + 11s + 10} = \frac{11s}{(s + 1)(s + 10)}$$

This has a real zero (at $s = 0$) and two real poles (at $s = -1$ and -10), so that breaks in direction occur at $\omega = 1$ and $\omega = 10$ rad/s. Plot a point on the first asymptote by picking an s value much smaller than this first breakpoint, let's say $s =$

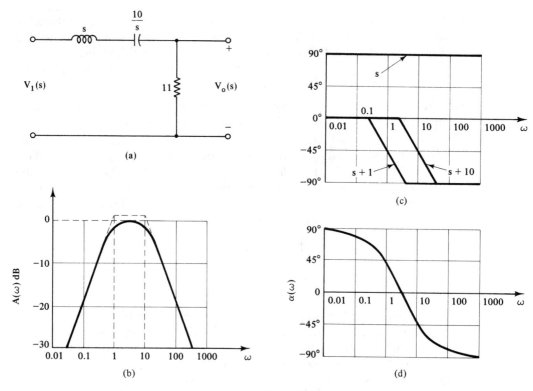

Figure E2.17

0.1. At this value, $G(s)$ is approximately equal to 0.11, or -19.1 dB. Plot this point. Since there is one zero at $s = 0$, sketch the first asymptote by sloping upward at 20 dB/decade from this plotted point. The first breakpoint occurs at $s = 1$. This is a simple pole, so the slope will change by -20 dB/decade. This means that the slope is now 0 dB/decade, or horizontal. At $s = 10$, another breakpoint occurs. It is again a simple pole, causing another -20 dB/decade change, so that the plot now bends down at -20 dB/decade. Figure E2.17b shows these three asymptotes and the true blended curve.

The phase plots are shown for the three factors in Figure E2.17c, and their sum is shown in Figure E2.17d. Notice that in regions where the gain is flat (horizontal), the phase is zero degrees, and the circuit is purely resistive ($X_C = X_L$).

Because the slopes of the gain plot asymptotes are straight lines we often refer to these slopes as **roll-off rates**. Thus in Example 2.17 we would say that the high-frequency roll-off rate is -20 dB/decade and the low-frequency roll-off rate is $+20$ dB/decade; and the break frequencies are 1 and 10 rad/s. A decade is a 10:1 change in frequency. In the case of gain being proportional to frequency, it is seen that a 10:1 change in frequency (a decade) produces a 10:1 change in gain. Converting 10 to decibels yields 20 dB. An octave is a 2:1 ratio. It has nothing to do with the number 8, except that in a 2:1 frequency range, musicians are able to fit the eight notes of a musical scale: do, re, mi, fa, sol, la, ti, do.   If gain doubles when frequency doubles (an octave), the gain in decibels is 6 dB. What is the roll-off rate if gain changes by 3:1 if the frequency changes by a ratio of 5:1?   Well, 5:1 is not one-half of a decade.   To find the number (or fraction) of decades, use the following formula:

$$\text{number of decades} = \log_{10} \frac{f_2}{f_1} \tag{2.30}$$

The gain change in decibels is given by the following formula:

$$\text{gain change (dB)} = 20 \log_{10} \frac{A_2}{A_1} \tag{2.31}$$

Thus the roll-off rate is found by dividing Eq. 2.31 by Eq. 2.30:

$$\text{roll-off rate} = 20 \frac{\log_{10}(A_2/A_1)}{\log_{10}(f_2/f_1)} \tag{2.32}$$

To convert decades to octaves, and vice versa, use this formula:

$$\text{number of octaves} = \frac{\text{number of decades}}{\log_{10}2} = \frac{\text{number of decades}}{0.301}$$

$$= 3.32 \times \text{number of decades}$$

The answer to the question is $20(\log_{10}3)/(\log_{10}5) = 13.6$ dB/decade.

Example 2.18

A circuit has a transfer function,

$$G(s) = \frac{0.08(s + 2.5)^2(s + 10{,}000)^2}{(s + 10)(s + 100)(s^2 + 200s + 100{,}000)}$$

Plot its gain response in decibels versus frequency.

Solution   First make a list of poles and zeros in ascending order.   Note that the two complex poles are at $-100 \pm j300$, so their magnitude is $\sqrt{100^2 + 300^2} = 316$. Therefore, the list is: $2.5(Z), 2.5(Z), 10(P), 100(P), 316(P), 316(P), 10{,}000(Z), 10{,}000(Z)$. Now pick a low s value, say $s = 1$.   Since all the s in $G(s)$ are combined with numbers greater than 1, the s can be ignored.   Therefore,

$$G(\text{low } s) = \frac{0.08(2.5)^2(10{,}000)^2}{10(100)(100{,}000)} = \frac{1}{2} = -6 \text{ dB}$$

Since there are no poles or zeros at $s = 0$, plot a horizontal asymptote at the -6-dB level over to $\omega = 2.5$, the first breakpoint.   Next, follow the list above by going up at 40 dB/decade to $\omega = 10$.   Since there are exactly two octaves from 2.5 to 10 (two doublings), this means a rise of 24 dB (because of the double zero at $s = 2.5$).   Then reduce the rate of climb by 20 dB/decade to 20 dB/decade until $\omega = 100$.   Then reduce the rate of climb by 20 dB/decade to 0 dB/decade starting at $\omega = 100$.   Then at the double pole at $s = 316$, drop at a rate of 40 dB/decade from $\omega = 316$ to $\omega = 10{,}000$,

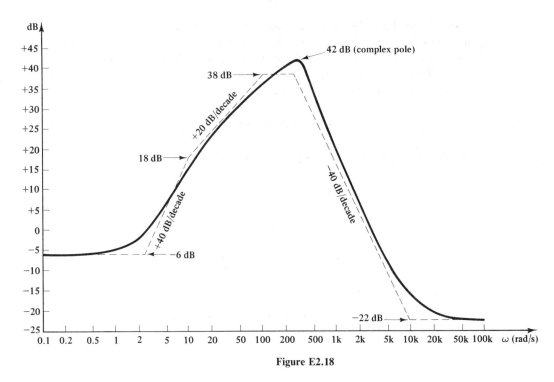

Figure E2.18

which is $1\frac{1}{2}$ decades. A final change of 40 dB/decade at the last zero will bring the slope to zero (horizontal). Now the blend of these asymptotes can be done, staying 3 dB away at the breakpoints, except 6 dB at the double zeros at $\omega = 2.5$ and 10,000. At the complex pole at $\omega = 316$, the gain must be calculated.

$$|G(j316)| = \frac{0.08(316)^2(10,000)^2}{316(316)\sqrt{(100,000 - 316^2)^2 + [200(316)]^2}}$$

$$= \frac{8 \times 10^{-2} \times 10^5 \times 10^8}{10^5 \sqrt{0 + 4 \times 10^9}} = \frac{8 \times 10^{11}}{6.33 \times 10^9} = 126 = 42 \text{ dB}$$

With this point also plotted, the blending can be completed, and the dB gain plot will look like Figure E2.18.

SUMMARY

In this chapter you learned how to apply Laplace transforms to study the behavior of various passive circuits. You began by applying the integral and differential theorems to solve integrodifferential equations. Then you noticed that equations such as KVL are actually this type of equation. This means that Laplace transforms can be applied to find voltages and currents in a circuit by KVL, mesh analysis, and so on. An alternative approach is to replace component values by their "transform impedance" and use this transform impedance in the circuit equations

(KVL, etc.) as if they were resistor values, except that initial conditions for capacitors or inductors had to be simulated by the artificial introduction of step or impulse sources into the circuit. The inverse Laplace transform of the solution is then done to get the time-domain solution.

Another useful concept, especially for circuits containing only one source, is transfer function theory. This states that the ratio of the output to the input, called $G(s)$, does not depend on what the input is, as long as everything is expressed in the s domain. Some of the uses for $G(s)$ are in finding the step response and impulse response of a circuit and for determining whether a circuit is stable or not. The stability of a circuit is also evident by constructing a pole–zero plot of $G(s)$.

Because series RLC and parallel RLC circuits are quite common, their response to step and impulse inputs was studied in detail and can be characterized in terms of rise time, overshoot, and settling time, as well as whether the circuit's output is overdamped, critically damped, or underdamped.

Finally, the most important area of study is the steady-state response of circuits. It was shown how to calculate the gain magnitude and phase equation if the transfer function was known. These quantities can be displayed on a Bode plot. The concept of decibels was introduced as the preferred way to plot this gain. Asymptotes were found helpful in doing a very quick sketch of these plots. The transfer function and Bode plots are used extensively in filter analysis, as you will notice in the coming chapters.

PROBLEMS

Sections 2.1 and 2.2

2.1. Consider the circuit in Figure P2.1.
    (a) Write a differential equation involving $i(t)$ and solve for it, using calculus.
    (b) Transform the differential equation above into the s-domain using O-2 and solve for $I(s)$.

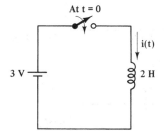

Figure P2.1

Section 2.3

2.2. (a) Redraw Figure P2.1 in the s-domain.
    (b) Solve the redrawn circuit for $I(s)$.
    (c) Find $i(t)$ by inverting $I(s)$.

2.3. **(a)** Transform Figure P2.3 into the *s*-domain.
  (b) Find $V_C(s)$ using VDR.
  (c) Find $v_C(t)$.

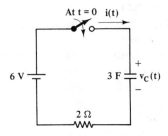

Figure P2.3

2.4. **(a)** Transform Figure P2.4 into the *s*-domain.
  (b) Find $I_L(s)$ using CDR.
  (c) Find $i_L(t)$.

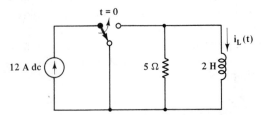

Figure P2.4

2.5. Find $v_C(t)$ for Figure P2.5.   (*Hint:* Use O-3).

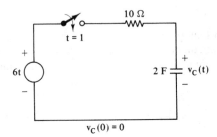

$v_C(0) = 0$ Figure P2.5

2.6. Find $i_L(t)$ for Figure P2.6.

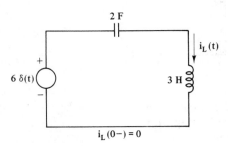

$i_L(0-) = 0$ Figure P2.6

2.7. (a) For Figure P2.7, find $v_C(t)$.

(b) Sketch $v_C(t)$ versus time, from $t = 0$ to $t = 24$ seconds.

2.8. If $v_C(0) = 7$ V in Problem 2.3:

(a) Redraw the circuit in the s-domain.

(b) Find $v_C(t)$.

(c) Find $i(t)$.

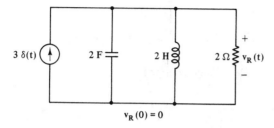

$v_R(0) = 0$ **Figure P2.7**

2.9. Redraw Figure P2.9 in the s-domain so that mesh analysis equations could be written (initial capacitor voltages and inductor currents are shown).

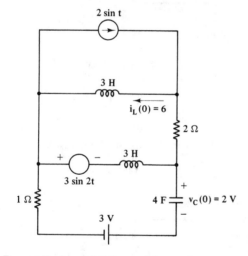

Figure P2.9

2.10. Repeat Problem 2.9 for nodal equations.

2.11. (a) Solve Figure P2.11 for $I_2(s)$ by mesh analysis.

(b) Find $i_2(t)$ by inverting $I_2(s)$.

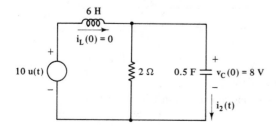

Figure P2.11

2.12. For the circuit in Figure P2.12, find $v_C(t)$ for $t > 0$.

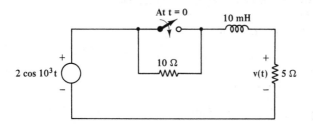

Figure P2.12

2.13. For the circuit in Figure P2.13, find $v_C(t)$ for $t > 0$.

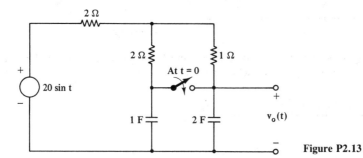

Figure P2.13

Section 2.4

2.14. $G(s)$ is $= V_o(s)/V_1(s)$ and $A = 1$ in the circuit of Figure P2.14.
   (a) Find $G(s)$.
   (b) Find $v_o(t)$ if $v_1(t)$ equals $5u(t)$.

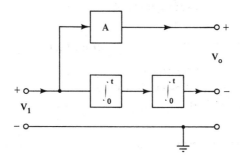

Figure P2.14

2.15. Repeat Problem 2.14 if $A = -1$.
2.16. Repeat Problem 2.14 if $A = -2$.
2.17. Find V_o/V_1 for the block diagram shown in Figure P2.17.

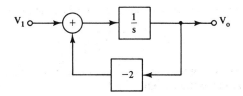

Figure P2.17

2.18. Using only adders, amplifiers, and integrators, set up a block diagram to have the following V_o/V_1 ratio:

$$\frac{V_o(s)}{V_1(s)} = \frac{s}{s + 2}$$

2.19. Find the driving-point impedance for Figure P2.19.

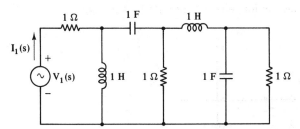

Figure P2.19

2.20. Find the transfer admittance, $I_2(s)/V_1(s)$ for Figure E2.10.

2.21. Draw the circuit whose driving-point impedance $= (s^2 + 1)/s$.

2.22. Synthesize (draw a schematic for) the circuit whose driving-point impedance is

$$\frac{288s^4 + 324s^3 + 276s^2 + 129s + 26}{48s^3 + 30s^2 + 27s + 6}$$

2.23. A circuit has an input $x(t) = te^{-t}$ and an output $y(t) = t^2e^{-t}$. If this $x(t)$ is replaced by a new $x(t) = 5u(t)$, what is the new $y(t)$?

2.24. In a circuit, $x(t) = 7 \sin t$ and $y(t) = 7 \cos t$. What does $x(t)$ have to be to produce $y(t) = 8e^{-2t}$?

Sections 2.5 and 2.6

2.25. A circuit's impulse response is $10e^{-3t}$.
  (a) What is the response to $\sin t$?
  (b) Synthesize the circuit if $G(s) = Z_{dp}(s)$.
  (c) What is the circuit's step response?

2.26. Find the $G(s)$ of a circuit whose output is $8u(t - 2)$ when its input is $8u(t)$.

2.27. Find $G(s)$ for a circuit whose step response is $3 \sin 5t + 2 \cos 5t$.

2.28. Find the response to a step of $6u(t)$ if a circuit's transfer function is

$$G(s) = \frac{4s^2 + 2s + 2}{s^2 + 2s + 2}$$

2.29. In a certain circuit, a 2-V dc input is switched on at $t = 0$, and the output is $e^{-2t}u(t)$. Find the input that will make the output $= 2u(t)$.

2.30. $(\sin t)u(t)$ volts is applied to the input of a circuit and the input current of the circuit is $3u(t)$ amperes.
(a) Find $Z_{in}(s)$.
(b) Synthesize the circuit.

Sections 2.7 and 2.9

2.31. From the pole–zero diagram for $G(s)$ in Figure P2.31, find $g(t)$.

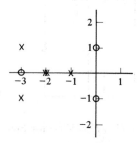

Figure P2.31

2.32. Find $G(s)$ in Problem 2.31 if $G(0) = 5$.

2.33. From the pole–zero plot of $G(s)$ in Figure P2.33, find $G(s)$ if $G(\infty) = 3/s^2$, that is, if $G(s)$ is approximately equal to $3/s^2$ as s gets very large.

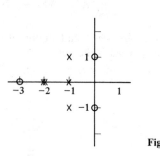

Figure P2.33

2.34. Characterize the following transfer functions, $G(s)$, as stable, unstable, or marginally stable.

(a) $\dfrac{(s - 2)(s + 3)}{s^2(s + 1)}$ **(b)** $\dfrac{(s + 1)(s + 3)}{(s^2 + 5)^2}$ **(c)** $\dfrac{s^2 + 4}{(s + 1)(s^2 + 1)}$

(d) Figure P2.31   **(e)** Figure P2.33

Sections 2.10 and 2.11

2.35. Find the time constant, rise time, overshoot, and settling time for the capacitor voltage in an RC series circuit with $R = 1\ k\Omega$ and $C = 100\ \mu F$ if the input is a step voltage.

2.36. Consider the circuit in Figure 2.8 with $L = 5\ mH$, $C = 100\ \mu F$, and $R = 10\Omega$.
  (a) Is it over, under, or critically damped?
  (b) What is the frequency?
  (c) If R were doubled, what would the circuit be?
  (d) What must R be for critical damping?

2.37. Repeat Problem 2.36 for Figure 2.10.

Section 2.12

2.38. A circuit with

$$G(s) = \frac{1}{(s + 2)^2(s + 1)}$$

is driven by a source whose transform is $1/(s^2 + 4)$. Find the steady-state response.

2.39. If

$$G(s) = \frac{6.4 \times 10^{10}}{(s + 4000)^3}$$

  (a) What frequency produces a $-45°$ phase shift?
  (b) What is the gain at this frequency?

Section 2.13

2.40. Make a Bode plot of $A(\omega)$ for Figure 2.8, showing asymptotes and actual curves. Use semilog paper and put $A(\omega)$ into decibels, using values from Problem 2.36.

2.41. Make a Bode plot of the circuit in Figure 2.10 if $L = 22\ H$, $R = 20\ \Omega$, and $C = 4545.45\ \mu F$. What is the transfer function, $G(s) = V_o/I_1$? Factor this function and plot $A(\omega)$, using asymptotes.

2.42. $G(s) = V_o/V_1$ for the circuit of Figure P2.42.

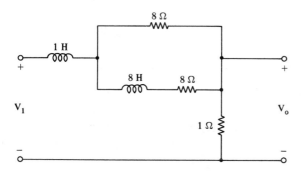

Figure P2.42

(a) Find $G(s)$ and factor it.

(b) Make an asymptotic Bode plot of the magnitude of $G(s)$ from 0.1 to 100 rad/s, or from 0.01 to 10 Hz.

(c) On this Bode plot, sketch the curve by using a 3-dB offset at the breakpoints.

(d) Use the NETWORK program to get an accurate curve (for frequency input, enter 0.01, 10, -100).

(e) Compare parts (c) and (d), especially on the disagreement in decibels at the breakpoints and your reasons.

3

analysis of passive filters

OBJECTIVES

After completing this chapter, you should be able to:

- Classify a filter by its technical field, passive/active, signal type, or frequency response by just glancing at the filter and its schematic.
- Recognize by the shape of the frequency response curve, whether it is a Butterworth, Chebyshev, inverse Chebyshev, or elliptic filter.
- Apply the procedures of network analysis (KVL, etc.) to a given circuit in order to find the $G(s)$ of that circuit.
- Determine the gain function $A(\omega)$, and the phase function, $\alpha(\omega)$, if the $G(s)$, of the filter is known.
- Examine the gain function, $A(\omega)$, to determine the order of the filter and whether it is Butterworth or not; and if so, what the cutoff frequency is.
- Examine the gain function, $A(\omega)$, to determine if it is a Chebyshev filter or not; and if so, find its ripple width in decibels, its order, and its cutoff frequency.

• Use the computer program to check a designed circuit to make sure that it meets the original spec.

INTRODUCTION

Most people think of filters as devices that cleanse the oil in an automobile or remove dust particles from the air as air passes through a hot-air furnace. Thus a filter is a sieve. A more general definition of a filter is a device that takes a mixture of things at the input and separates some of them from the rest. The reason this is done is because some of these things may be harmful to equipment; in other words, they are unwanted.

In the context of our studies, we are not concerned with material particles but with signals. In the electromechanical field, these may be mechanical vibrations at various frequencies or electrical voltages or currents at certain frequencies. The unwanted signals may be harmful to our equipment, or more usually, they just get in the way of the signal in which we are really interested. Since we are talking about sinusoidal frequencies and not about transient behavior, the subject of filters follows nicely the last two sections. In other words, the frequency plots developed in the last two sections give a very good picture of which frequencies pass most easily through our "sieve" and which ones do not pass, or pass through greatly diminished in magnitude. For this reason, Bode plots are used quite extensively in this chapter.

3.1 CLASSIFICATION OF FILTERS

There are many different ways of classifying filters. One basis might be the physical phenomena involved. For instance, a mechanical filter would be one that seeks to modify mechanical vibrations. Let's take the case of a highway. No highway is as smooth as glass. There are small bumps and ripples in even a brand-new highway. If we rode a bicycle with no tire on its rim, we would feel these little ripples and the ride would be quite uncomfortable. Attaching a solid rubber tire to the rim removes these small high-frequency (because the ripples are closely spaced) ripples. Attaching an air-filled tire does even a better job; it even prevents some of the potholes from affecting us full strength. So the tire can be considered as a mechanical filter. Automobiles require even more filtering than can be achieved with their big tires; they use springs and shock absorbers to filter or reduce the magnitude of large potholes.

In the field of hydraulics the ocean can be considered as a low-pass filter, since low-frequency waves such as large swells or tidal waves can travel great distances; but small, closely spaced ripples quickly die out.

In the field of acoustics, a thin-walled wine glass is a narrow bandpass filter, allowing only a small range of high frequencies to affect the glass and set it vibrating.

When a powerful voice sings at the right frequency and close enough to the fragile glass, it could shatter.

In the electronic field, we are concerned mostly with voltages and currents and components such as resistors, inductors, capacitors, and amplifiers. However, there is some overlap with the other branches of science. Transducers are devices that connect or convert signals from one branch of science to another. For instance, an oil pressure sensor (a pressure-to-electrical transducer) in an automobile will convert hydraulic pressure to an electrical voltage, which then causes the dashboard gauge pointer to move. Thus the dashboard gauge is an electrical-to-mechanical transducer. There is an endless variety of these transducers found in almost every piece of equipment we see. The field of robotics is certainly rich in many of these transducers. Because of this, it is important that someone who specializes in electronics have some knowledge of these other fields, and also that anyone trained in the mechanical engineering field have some knowledge of electronics. This is one of the biggest reasons why the study of physics is so important for anyone in a technical field, since it ties many technical fields together.

In the case of a car's suspension system as shown in Figure 3.1, if force is analogous to voltage and if velocity is analogous to current, it is seen that the shock absorber is in series with the spring, since one end moves with respect to the other end with the same velocity for both components, so the "current" is the same. This series combination is in parallel with the mass, since the mass has the same force as the sum of the two other forces, but the velocity "through" it is the velocity relative to ground. This means that masses always have one terminal grounded. The mathematical equations for this system look so much like that for an electrical circuit that an analogy can be constructed. Springs are capacitors, friction elements such as shock absorbers are resistors, masses are inductors, the vertical velocity of

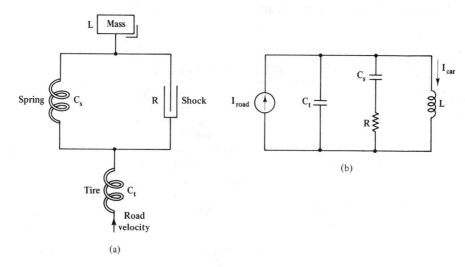

Figure 3.1

the road's surface is an ac current source, and the force on any component is its voltage. What this all means is that all the rules of circuit analysis and transform analysis can be used in other fields to great benefit.

Besides the study of electrical filters, the most important filters for an electronics technologist to study are those that involve electrical signals and signals from another field of study. These other elements are used because they are perhaps cheaper, more accurate, or take up less room. Thus one logical way of classifying filters would be: purely electrical versus a combination of electrical and some other field, such as mechanical, usually involving some kind of transducer.

Another way of classifying filters is by specifying if they are analog or digital. Analog filters work with continuous-time signals, such as we have discussed up until now. Digital filters work with discrete-time signals; that is, only samples of the original input are used, perhaps spaced at equal time intervals.

Another way of classifying filters is by specifying if they are passive or active. Most of our work has been with passive filters. Active filters involve putting amplifiers into our circuit, usually with feedback loops. The danger is that unstable circuits can result if the circuit is not designed properly. However, the benefit is that a larger variety of useful circuits can be created.

Finally, another way of classifying filters is by the shape of their gain response or phase response versus frequency. The names of some of these are: low-pass, high-pass, bandpass, band-reject (or notch filter), all-pass, linear phase, and lead-lag. There are several ways of implementing these filters: Butterworth, Chebyshev, Bessel, elliptic.

Because of all these various ways of classifying filters, we can specify a filter by all these methods at the same time. For instance, one could specify a bandpass Butterworth active analog quartz filter, or a low-pass elliptic active digital electric filter. The possibilities seem almost endless. Table 3.1 summarizes the different ways that filters can be classified.

TABLE 3.1 FILTER CLASSIFICATIONS

Characteristics	Classifications
Physical nature	Electrical, electromechanical, electroacoustic
Type of element	Passive, active
Type of signal	Analog, digital
Type of response	Low-pass, high-pass, bandpass, band-reject, all-pass, uniform phase shift
Details of shape of response	Butterworth, Chebyshev, inverse Chebyshev, elliptic, Bessel
Complexity	First order, second order, third order, etc.

3.2 APPLICATIONS OF FILTERS

An application of a passive analog low-pass purely mechanical filter was the automobile suspension system discussed in Section 3.1. The following cases are examples of some filters that are at least partly electrical in nature.

A good example is the timing circuit in a digital wristwatch. For accuracy, an oscillator is needed which has a high Q, and is very stable; that is, its frequency does not change with temperature, humidity, time, and so on. A quartz crystal, whose symbol is shown in Figure 3.2, is such a device. If a quartz crystal, which is nonconductive, is plated on opposite surfaces, it behaves as a small-valued capacitor C_e. However, when voltage is applied, it has been discovered that the force of attraction between the opposite charges on these surfaces will compress the glasslike quartz material a tiny bit. If the voltage is ac, the effect is to cause the quartz to vibrate. This means that the mass and the springiness of the crystal are behaving analogous to an inductor, L_m, and a capacitor, C_s. The analogous circuit is shown in Figure 3.2b, where R_d has been added to show the mechanical damping of this crystal. All materials have damping; for instance, if a tuning fork is struck, its sound will gradually diminish. It turns out that for the quartz crystal, R_d is very small and its resonance is very sharp. This makes it an excellent choice for this application. It is also rugged, cheap, and small.

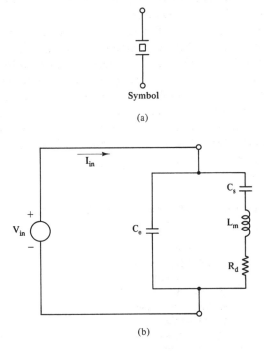

Symbol

(a)

(b)

Figure 3.2

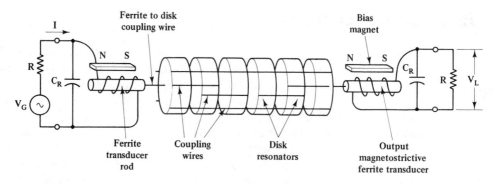

Figure 3.3
Reprinted by permission from Robert A. Johnson, "Mechanical Filters in Electronics," John Wiley & Sons, Inc., copyright, 1983.

An example of an electromechanical filter is shown in Figure 3.3. Its input and output are electrical, but in the middle it is mechanical. This means that a transducer is required at both ends, and a series of rods and disks between these transducers vibrate in such a manner as to provide the required frequency response. When an accurate filter is required that has a sharp cutoff and needs many elements, it can usually be realized most cheaply and in the smallest space, by using this type of filter. They are used in such applications as single-sideband communications filters, and in demultiplexing of phone signals.

Digital filters are almost always purely electrical in nature. They involve a computer or a microprocessor as part of their circuit. They may be constructed with digital hardware or with digital software. A sampling circuit, called an **analog-to-digital converter** (A-D converter), converts an input that is a continuous function of time into discrete samples. These samples are then processed by the computer. The output of the computer is a series of discrete values which are then put through a simple low-pass analog filter (called a D-A converter), so that the final output is once again a continuous function of time. If the input and output are digital in nature, the two converters are not necessary.

Example 3.1

Show that the following BASIC program is a high-pass digital filter, done in software.

```
10 V2 = 0
20 REM Get a sample
   of the input
30 READ Vin
40 Vo = V2 − Vin
50 PRINT Vo
60 V2 = Vin
70 GO TO 30
```

Solution   Let us assume that samples are taken every millisecond and let us see what happens when the input is at 250 Hz and then at 41.7 Hz. The following table shows the voltage, V_{in}, generated by the sampling circuit (not shown) and the output voltage, V_o, generated by the BASIC program.

						t (ms)					
	0	1	2	3	4	5	6	7	8	9	10
250 Hz											
V_{in}	0	1	0	−1	0	1	0	−1	0	1	0
V_o	0	−1	1	1	−1	−1	1	1	−1	−1	1
41.7 Hz											
V_{in}	0	0.26	0.50	0.71	0.87	0.97	1.0	0.97	0.87	0.71	0.50
V_o	0	−0.26	−0.24	−0.21	−0.16	−0.10	−0.03	0.03	0.10	0.16	0.21

This shows clearly that for the high frequency the output has a peak value of 1.0, and for the lower frequency the output has a peak of only 0.26; indicating this is probably a high-pass filter. Several more frequencies would have to be tried to tell for sure that this is indeed a high-pass filter.

Digital filters find application in compact disk recording technology, computer enhancement of photographs taken from outer space, and other filter applications where digital technology is replacing analog circuitry, such as in automobiles (trip computers, miles/gallon calculations, antiskid braking devices). Instead of Laplace transforms, z-transforms are very useful in analyzing digital filters. This is covered in Chapter 8.

The bulk of this book is concerned with analog purely electrical filters, with Chapters 3 and 4 being devoted to passive filters. Before the advent of op-amps and microprocessors, all filters were of the passive analog type. Examples include the crossover networks used in high-fidelity speaker systems, the treble and bass controls of high-fidelity equipment, equalization networks in phonograph systems, preemphasis circuits in FM broadcasts, and tank circuits in a radio's tuner.

Example 3.2

Show that the circuit in Figure E3.2 is an all-pass filter if $G(s) = V_o(s)/V_1(s)$.

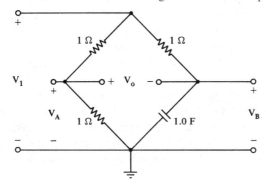

Figure E3.2

Solution   Converting the circuit to the s-domain and using VDR on each branch:

$$V_A = \frac{V_1}{2}$$

$$V_B = \frac{V_1(1/s)}{1 + 1/s} = \frac{V_1}{s + 1}$$

$$V_o = V_A - V_B = \frac{V_1(\frac{1}{2})(s - 1)}{s + 1}$$

$$G(s) = \frac{(\frac{1}{2})(s - 1)}{s + 1}$$

$$G(j\omega) = \frac{1}{2} \frac{\sqrt{1 + \omega^2} \; \angle \tan^{-1}(\omega/-1)}{\sqrt{1 + \omega^2} \; \angle \tan^{-1}(\omega/1)}$$

$$= 0.5 \; \angle -2 \tan^{-1}(\omega) - 180°$$

Thus the magnitude does not change with frequency, although the phase does.

Almost all of the passive filters mentioned above could be realized in an active form. The advantage is amplification of the signal, high input impedance so as not to load the input, low output impedance so as not to have the filter be loaded by any circuitry that follows, the avoidance of inductors, and the absence of interaction, or loading, between sections of the circuit. This means that the $G(s)$ can be broken up into several sections, with each section performing the function of one or more poles and/or zeros of $G(s)$. Chapters 5 and 6 deal with active filters.

The concept of switched capacitors has come along recently. It is useful in the design of active filters, because it eliminates resistors from the filter circuit, leaving capacitors as the only passive elements. The advantage of this step is an increase in accuracy of components, as we shall see in Chapter 7. This type of circuit is sometimes called an analog sampled data system, because the input is sampled at equal time intervals just as in a digital system, but then these sample pulses are processed in what amounts to an analog circuit. Switched capacitor modules are available in integrated-circuit form. By adding a few external components, any type of frequency response can be achieved using this standard module. Chapter 7 is an introduction to switched capacitors. It is also possible to use switched capacitors to simulate a delay. If such a delay unit is incorporated into a switched capacitor circuit, it is possible to get a "switched capacitor" version of a digital filter.

All the filters we have discussed are lumped parameter, time-invariant, linear filters. The opposite of lumped is distributed, meaning that resistance, inductance, and capacitance are bound together such that they cannot be separated into individual components for purposes of analysis. A time-varying system is one whose R, L, or C varies with time. A nonlinear system is one in which R, L, or C vary with voltage or current. For instance, a diode has a "resistance" that drops very quickly as more voltage is applied. There are some capacitors whose capacitance varies with voltage; these are called **varactors** and are useful in frequency-modulation generation. None of these types of circuits will be considered here. Analysis of such circuits is usually quite complex and their study belongs more properly in a graduate-level course.

3.3 GAIN AND PHASE FROM THE FILTER TRANSFER FUNCTION

The performance of a filter is mathematically best described by its transfer function, $G(s)$, which is the ratio of the output, $Y(s)$, to the input, $X(s)$. This was discussed in detail in Section 2.4. Input and output are usually defined to be voltages, although this is not strictly necessary.

For instance, for the circuit of Figure 3.4,

$$\frac{V_{out}(s)}{V_{in}(s)} = \frac{10^6/s}{100 + 10^6/s} = \frac{10,000}{s + 10,000} = G(s) \tag{3.1}$$

This is also recognized as a low-pass filter since we see that at low frequencies the capacitor's impedance is higher than the resistor, hence almost all the voltage appears across the capacitor. But at high frequencies, the capacitive impedance is smaller than the resistor and only a small voltage appears across the capacitor, since most of the voltage drop is across the resistor. By looking at the circuit in this manner, it is obvious that this is a low-pass filter.

However, by looking at the mathematical expression for $G(s)$, this is not as obvious. Two considerations are important. First, the concept of low-pass or high-pass implies that we are considering sinusoidal inputs only. Second, if we want to decide whether we have a high-pass or a low-pass, and so on, we must do calculations at various frequencies, so that a table of output-to-input ratios versus frequency would be set up, so that we could compare results. Actually, the best way to compare results would be to plot the output-to-input ratio versus frequency. Visual examination of this "frequency plot" would immediately tell us what type of filter we have.

For the first consideration, in which we are to be dealing with sinusoidal inputs, we can treat this as a problem in ac analysis, using phasors. Using the

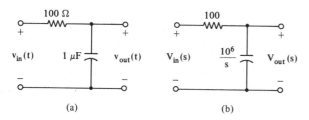

(a)   (b)   **Figure 3.4**

voltage-divider rule on Figure 3.4a, we have

$$V_{out} = V_{in} \frac{-j[1/(\omega 10^{-6})]}{100 - j[1/(\omega 10^{-6})]}$$

or

$$\frac{V_{out}}{V_{in}} = \frac{10,000}{10,000 + j\omega} \qquad (3.2)$$

Notice that this is similar to the Laplace notation for $G(s)$, except that $j\omega$ appears where the s used to be. This is true for all circuits; that is, the phasor result for sinusoidal excitation is

$$\frac{V_{out}}{V_{in}} = G(j\omega)$$

Another way in which we can apply $G(s)$ to the sine-wave case is to let $v_{in}(t) = \sin \omega t$, or $V_{in}(s) = \omega/(s^2 + \omega^2)$. Then

$$V_{out}(s) = G(s)V_{in}(s) \qquad (3.3)$$

$$= \frac{10,000}{s + 10,000} \frac{\omega}{s^2 + \omega^2} = \frac{A}{s + 10,000} + \frac{Bs + C}{s^2 + \omega^2}$$

Equation 3.3 is due to partial fraction expansion. From Section 2.8 we see that the character of the time expression is due to the denominators of the partial fractions. Since the first one is an exponential, it is transient in nature and is of no interest to us in this section. The second one is the steady-state sinusoidal result. The frequency is that of the source and the magnitude depends on the source and on $G(s)$. The inverse Laplace of this term is determined most easily by using the "trick" formula of Section 1.11C. For the second term, $\alpha = 0$, $\omega = \omega$, and $Q(s) = 10^4\omega/(s + 10^4)$. Therefore,

$$Q(j\omega) = \frac{10^4\omega}{s + 10^4}\bigg|_{s=j\omega} = \frac{10^4\omega}{j\omega + 10^4} = \frac{10^4\omega}{\sqrt{\omega^2 + 10^8}} \bigg/ \tan^{-1}\frac{\omega}{10^4} = M \angle \theta$$

or

$$f(t) = \frac{M}{\omega}e^{-\alpha t}\sin(\omega t + \theta) = \frac{10^4}{\sqrt{10^8 + \omega^2}} \bigg/ \tan^{-1}\frac{\omega}{10^4} \qquad \text{phasor form} \qquad (3.4)$$

We see that $f(t)$ is given in terms of a magnitude and a phase, both depending on the frequency, ω, and these expressions agree with Eq. 3.2, after converting that equation to polar form. Thus one method confirms the other. Thus for sinusoidal inputs, the ratio of output phasor to input phasor is $G(j\omega)$.

For the second consideration, how do we best represent $G(j\omega)$ so as to visualize what type of filter we have? The best approach is to convert $G(j\omega)$ to polar form, as we did in the preceding paragraph; that is,

$$G(j\omega) = A(\omega) \, \underline{/\alpha(\omega)} \tag{3.5}$$

$A(\omega)$ is sometimes written as $|G(j\omega)|$, where the absolute sign means the $G(j\omega)$ magnitude only. Also, $\alpha(\omega)$ is sometimes written as $\underline{/G(j\omega)}$, which means the $G(j\omega)$ angle only.

Either $A(\omega)$ or $\alpha(\omega)$ are plottable functions. Consider the $G(j\omega)$ of Eq. 3.2, which is shown in polar form in Eq. 3.4. We see

$$A(\omega) = \frac{10^4}{\sqrt{10^8 + \omega^2}} \quad \text{and} \quad \alpha(\omega) = -\tan^{-1}\frac{\omega}{10^4} \tag{3.6}$$

Plots of $A(\omega)$ and $\alpha(\omega)$ versus frequency are shown in Figure 3.5a and b. From the figure it is obvious that this is a low-pass filter.

For the special case of making the V_{in} equal to unity for all frequencies, $A(\omega)$ will be identical to the magnitude of V_{out}. This is a common practice in the lab when we are testing a filter; that is, we will keep the input voltage constant but vary its frequency and note the effect this has on the magnitude of the output of the filter. The phase function is the difference between output and input phase angles.

Example 3.3

For the circuit shown in Figure E3.3a, what type of filter is it, and what is the cutoff frequency (frequency at which V_{out} is 70.7% of the maximum graphed value)?

Solution   By VDR,

$$G(s) = \frac{s(0.01)}{s(0.01) + 500} = \frac{s}{s + 5 \times 10^4}$$

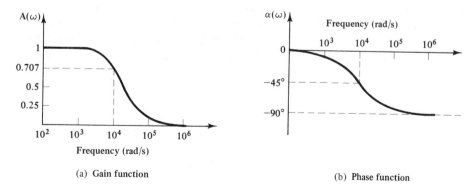

(a) Gain function

(b) Phase function

Figure 3.5

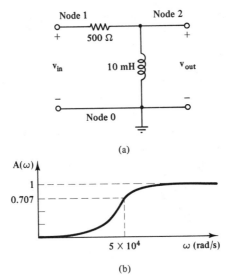

(a)

(b)   **Figure E3.3**

Therefore,

$$G(j\omega) = \frac{\omega}{\sqrt{\omega^2 + 25 \times 10^8}} \bigg/ \!\!\!\! \underline{90° - \tan^{-1}\left(\frac{\omega}{5 \times 10^4}\right)} = A(\omega) \, \underline{/\alpha(\omega)}$$

A plot of $A(\omega)$ is shown in Figure E3.3b, from which it is obvious that this is a high-pass filter whose cutoff is at 5×10^4 rad/s, which is the same as 8 kHz.

3.4 COMPUTER ANALYSIS OF FREQUENCY RESPONSE

To avoid having to calculate and plot all these points by hand, we can use the BASIC program called NETWORK, which is given in Appendix B. To use this program, the input and output must be connected to a common reference node (called "zero") and the input must be a voltage source. We must number all the nodes (see Figure E3.3a). This circuit is called a two-node circuit. The network data are entered as follows:

R

1,2,500

L

0,2,1E-2

E

1,2

This last entry signifies the end of the data entry and also that the input is at node 1 and the output is at node 2. Then the program will ask for a range of frequencies to be plotted and the spacing between frequencies. If the results look uninteresting, the program gives us the opportunity to change the frequency range (we want to be sure to include the cutoff frequency in the range of the results). For instance, if we enter (in hertz)

$$1600,16000,400$$

we will cover the range from 1.6 to 16 kHz with 400-Hz spacing. This will give us 37 data points and will only take about 2 minutes to run. However, the program cannot handle more than 100 frequencies. The output will be given in decibels, not in volts. The decibel value is equal to $20 \times \log_{10}(V_{out}/V_{in})$. If we plot this ourselves or ask the computer to plot this on the screen or on the printer, we will get a very smooth high-pass filter.

Example 3.4

For the circuit shown in Figure E3.4a, find the type of filter, its cutoff frequency, and center frequency.

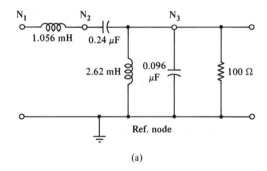

(a)

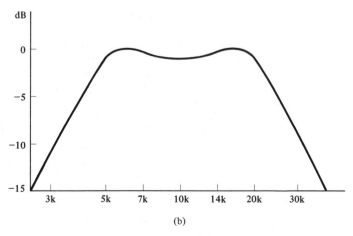

(b)

Figure E3.4

Solution   By converting the circuit to the s-domain and using VDR, we could calculate the $G(s)$.   It would turn out to be

$$G(s) = \frac{0.98 \times 10^{10}s^2}{s^4 + 1.035 \times 10^5s^3 + 1.77 \times 10^{10}s^2 + 0.409 \times 10^{15}s + 0.156 \times 10^{20}}$$

When $j\omega$ is substituted for s and the expression is converted to polar form, the magnitude portion of the result is

$$A(\omega) = \frac{0.98 \times 10^{10}\omega^2}{\sqrt{\omega^4 - 1.77 \times 10^{10}\omega^2 + 0.156 \times 10^{20})^2 + (-1.035 \times 10^5\omega^3 + 0.409 \times 10^{15}\omega)^2}}$$

$A(\omega)$ could be calculated for several frequencies.   This would be made easier if a programmable calculator were available.   However, it is even easier if we use the NETWORK program.   In that case, we do not even have to find the two equations above.   In that case, the program is capable of printing out the frequency plot shown in Figure E3.4b.   Two further pointers in the use of this program follow:

1. To enter capacitors, enter their microfarad values.   Thus if it is a 1.5-μF capacitor, enter 1.5, not 1.5E-6.
2. The frequency range selected for this example was 2.5 to 40 kHz.   For such a large ratio, a uniform spacing of frequencies is inappropriate.   For more even spacing on log paper, a logarithmic spacing is desired; that is, the ratio of any frequency to the previous frequency should be the same no matter what portion of the graph we are at.   Instead of specifying the spacing in hertz, we specify the number of points we want, preceded by a minus.   In this problem the specification (81 points) is

$$2500, 40000, -81$$

By looking at this graph, we conclude that this is a bandpass filter with a center frequency of 10 kHz, a 3-dB lower cutoff of 4400 Hz, an upper cutoff of 23,700 Hz, and a 3-dB bandwidth of 19,300 Hz.   However, other decibel values could be used as the cutoff criterion.   Since this filter has an up-and-down motion in the passband (called the passband ripple) of 1 dB, it is customary to specify the cutoff frequency at the decibels of this ripple.   This would mean that the 1-dB frequencies are 5 and 20 kHz and the 1-dB BW is 15 kHz.   A filter that has this type of ripple in the passband is called a Chebyshev filter, to be discussed later.   Notice that just by looking at the circuit we could have told that it was a bandpass filter, because at low frequencies the 0.24 μF capacitor would have blocked most of the signal and the 2.62-mH inductor would have effectively shorted the load, and the series inductor and the shunt capacitor would have done the same thing at high frequencies.   However, at 10 kHz, the series components would constitute a short and the shunt components an open, thus allowing full voltage at the load.   The resonant condition that causes the voltage maximums (1 dB higher) above the "full" voltage could not be predicted by just "looking" at the circuit, or even by doing a Bode analysis (discussed later).   So this example is a case where the computer program is a very valuable tool to verify that a given circuit will perform as specified.

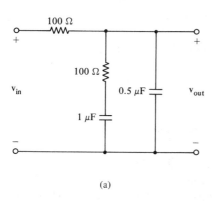

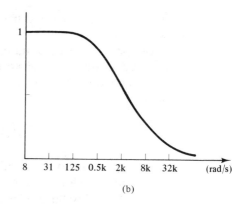

(a)   (b)

Figure 3.6

3.5 BUTTERWORTH FILTER

Consider the LP (low-pass) filter shown in Figure 3.6. The response curve decreases monotonically toward zero as the frequency increases. The word "monotonic" means that the slope is always negative (for LP) or it is always positive (for HP). Although this filter may be useful for some particular application, it does not have a name. In fact, there are countless specialized filters that have no name.

The ideal LP filter is one that passes all low frequencies equally well and does not pass any of the higher frequencies. This type of filter does have a name. It is called the brick-wall filter, so-named because of the shape of its frequency response curve (see Figure 3.7). This is the ideal LP filter shape, yet it is unattainable in practice. However, there are a number of real filters that come close to this ideal. In fact, if we add more components of the proper value, the ap-

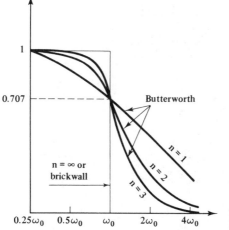

Figure 3.7

proximation gets better and better. In this section we consider the Butterworth approximation.

Butterworth filters of different order of complexity (order = n) are graphed in Figure 3.7 and compared to the brick-wall filter. The corresponding circuits and gain equations are shown in Figure 3.8. The general equation for this type of filter is

$$A(\omega) = \frac{k}{\sqrt{1 + (\omega/\omega_0)^{2n}}} \tag{3.7}$$

where k is an arbitrary constant and ω_0 is the 3-dB cutoff frequency in rad/s.

The order, n, for a LP filter is also the number of reactive elements (capacitors and inductors). From Figure 3.7 it seems that as n increases, the shape approaches the brick-wall shape. However, it would take an infinite number of zero-tolerance capacitors and inductors to achieve the perfect brick-wall filter.

Thus, if we are given the circuit, we can find the gain equation. Then, if this equation looks like Eq. 3.7, it is Butterworth and, when plotted, will have the typical Butterworth shape. This filter is sometimes also called the **maximally flat filter**, since the first n derivatives of the gain function are zero at $\omega = 0$ (for the LP case).

The criterion for cutoff is arbitrary, but it is customary to specify that frequency at which the gain function is 70.7% (i.e., 3 dB down) of the maximum

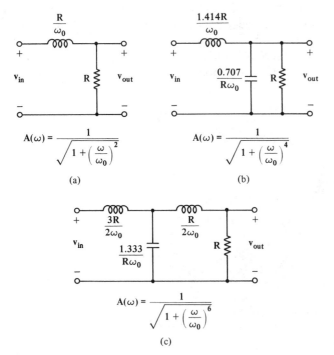

Figure 3.8

value.   The bandwidth is the width of the region of the curve (in rad/sec or hertz) where the gain function is above cutoff.   In Figure 3.7 this was $\omega_0 - 0 = \omega_0$ rad/s.

Example 3.5

Show that the circuit in Figure E3.5 is a Butterworth and find its cutoff frequency and order.

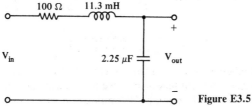

Figure E3.5

Solution   By VDR:

$$G(s) = \frac{1/sC}{R + sL + 1/sC} = \frac{1}{s^2LC + sRC + 1}$$

$$|G(j\omega)| = \left| \frac{1}{1 - \omega^2LC + j\omega RC} \right| = \frac{1}{\sqrt{(1 - LC\omega^2)^2 + R^2C^2\omega^2}}$$

$$= \frac{1}{\sqrt{(1 - 2.54 \times 10^{-8}\omega^2)^2 + (2.25 \times 10^{-4}\omega)^2}}$$

$$= \frac{1}{\sqrt{1 + 6.43 \times 10^{-16}\omega^4}} = \frac{1}{\sqrt{1 + (\omega/6280)^4}}$$

Comparing this to Eq. 3.7, we see that it is Butterworth and the cutoff is 6280 rad/s (or 1000 Hz).   Since the $2n$ from Eq. 3.7 is 4 in this case, the order, n, must be 2.   This is also confirmed by the fact that there are two reactive components.

As the next example shows, the 3-dB cutoff frequency occurs when the value under the radical of the denominator becomes twice what it would be at the extreme value of gain.

Example 3.6

For the circuit in Figure E3.6, find out if it is Butterworth and find its order and cutoff frequency.

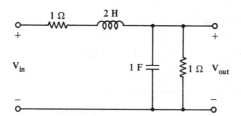

Figure E3.6

Solution   By VDR,

$$G(s) = \frac{1(1/s)/[1 + 1/s]}{[1(1/s)/(1 + 1s)] + 2s + 1} = \frac{1}{2s^2 + 3s + 2}$$

Then

$$A(\omega) = |G(j\omega)| = \left|\frac{1}{(2 - 2\omega^2) + j\omega 3}\right| = \frac{1}{\sqrt{(2 - 2\omega^2)^2 + (3\omega)^2}}$$

$$= \frac{1}{\sqrt{4 + \omega^2 + 4\omega^4}}$$

Since the highest power under the radical is 4, this is a second-order LP filter.   However, the filter is not Butterworth, since the radical contains an extra term.   Also for this reason, the cutoff cannot be determined directly.   Instead, we note that the gain at $\omega = 0$ is $1/\sqrt{4 + 0 + 0} = \frac{1}{2}$.   Examination of the expression shows that this is the maximum gain, since any value of ω greater than zero will increase the denominator and decrease the gain.   Therefore, for this example, cutoff will occur when the gain $= 0.707(\frac{1}{2})$, or $1/(2\sqrt{2})$.   This means that the denominator, $\sqrt{4 + \omega_0 + 4\omega_0^2} = 2\sqrt{2}$, or

$$4 + \omega_0 + 4\omega_0^2 = 8 \quad \text{or} \quad 4\omega_0^2 + \omega_0 - 4 = 0$$

Solving this quadratic yields

$$\omega_0 = 0.88 \text{ rad/s}$$

Of course, another way of determining cutoff is just to plot the function and see at what frequency the gain drops to 70.7%.

Examples 3.5 and 3.6 were both second-order filters, but Example 3.5 was Butterworth, while Example 3.6 was not (in fact, it has no name).   In fact, any filter with an order higher than 1 may or may not be a Butterworth filter, whereas a first-order filter must be Butterworth.

Many students believe that a second-order Butterworth filter can be built by just tacking two first-order filters together.   The following two examples show that this cannot be done.

Example 3.7

Find out if the filter shown in Figure E3.7 is Butterworth.

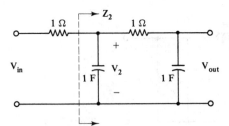

Figure E3.7

Solution   The circuit consists of two first-order Butterworth filters in cascade.   Since there are two capacitors, it is a second-order filter.   Applying VDR twice, we can get the transfer function, $G(s)$:

$$\frac{V_{out}}{V_2} = \frac{1/s}{1/s + 1} = \frac{1}{s + 1} \qquad Z_2 = \frac{(1/s + 1)(1/s)}{1/s + 1 + 1/s} = \frac{s + 1}{s(s + 2)}$$

$$\frac{V_2}{V_{in}} = \frac{Z_2}{Z_2 + 1} = \frac{s + 1}{s^2 + 3s + 1}$$

Therefore,

$$G(s) = \frac{V_{out}}{V_{in}} = \frac{V_{out}}{V_2}\frac{V_2}{V_{in}} = \frac{1}{s + 1}\frac{s + 1}{s^2 + 3s + 1} = \frac{1}{s^2 + 3s + 1}$$

From this, the gain magnitude equation is

$$A(\omega) = \frac{1}{\sqrt{1 + 7\omega^2 + \omega^4}}$$

This is not Butterworth, since it contains a center term.   At small and large ω, the gain is 1 and $1/\omega^2$, which is the same as for the Butterworth (from Eq. 3.7, with $\omega_0 = 1$). But at $\omega = 1$, the gain is $\frac{1}{3}$, which is about -9.6 dB, which is a far cry from Butterworth's -3-dB figure.

One of the reasons for the poor performance of the filter above (called an *RCRC* filter) is that the second section loads the first section.   The following example tries to overcome this difficulty by using a higher impedance level in the second section.

Example 3.8

Repeat Example 3.7 for the circuit in Figure E3.8.

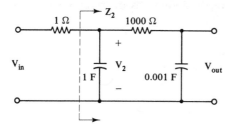

Figure E3.8

Solution

$$\frac{V_{out}}{V_2} = \frac{10^3/s}{10^3/s + 10^3} = \frac{1}{s + 1} \quad \text{(as before)}$$

$$Z_2 = \frac{(10^3/s + 10^3)(1/s)}{10^3/s + 10^3 + 1/s} = \frac{1}{s} \quad \text{(approx.)}$$

$$\frac{V_2}{V_{in}} = \frac{Z_2}{Z_2 + 1} = \frac{1/s}{1/s + 1} = \frac{1}{s + 1}$$

so

$$G(s) = \frac{V_{\text{out}}}{V_{\text{in}}} = \frac{V_{\text{out}}}{V_2} \frac{V_2}{V_{\text{in}}} = \frac{1}{s+1} \frac{1}{s+1} = \frac{1}{(s^2 + 2s + 1)}$$

and

$$A(\omega) = \frac{1}{\sqrt{(1 - \omega^2)^2 + (2\omega)^2}} = \frac{1}{\sqrt{1 + 2\omega^2 + \omega^4}} = \frac{1}{1 + \omega^2}$$

This is not of the same form as Eq. 3.7, so this is also not Butterworth, although it behaves like the second-order Butterworth function at very low and very high frequencies. However, at $\omega = 1$, the gain is $\frac{1}{2}$, which is down 6 dB. This is an improvement over the *RCRC* filter of Example 3.7, but is still 3 dB worse than the Butterworth function at $\omega = 1$.

3.6 CHEBYSHEV FILTER

Of the non-Butterworth filters of order greater than 1, there are three types that have some interesting properties and therefore carry a name. One of these is called a **Chebyshev filter**. The LP version is nonmonotonic; that is, in the passband the gain goes up and down as the frequency increases (see Figure 3.9). This figure shows a 3-dB variation in the passband, just as did the Butterworth, although we will see later that we can make this amount of variation anything we like by just picking the proper component values. But since the second-order Chebyshev filter we have picked has a 3-dB variation, it is easy to compare it with the second-order Butterworth filter, which also has a 3-dB variation in the passband.

Since the passband has the same variation (even though the curves have different shapes), they are considered to be of equal "goodness" in the passband. But in the stopband, arbitrarily defined as everything above ω_s, we see that the

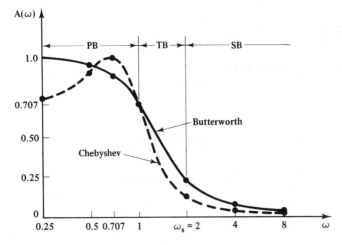

Figure 3.9

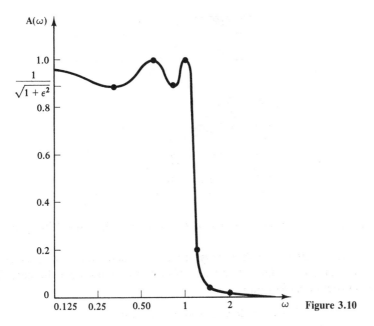

Figure 3.10

Chebyshev is always better; that is, it has a lower gain than the Butterworth. At higher frequencies it is about 6 dB lower (for second order) than the Butterworth. The circuit is no more complex than the Butterworth. In fact, it is identical, except that the components have different values. The price we pay for this improvement is the up-and-down motion in the passband, called **passband ripple**. This is considered to be a very small price.

The number of changes in direction of the curve in the passband plus one is called the number of half-ripples, and for the Chebyshev LP or HP filter it is the same as the order of the filter, which is also the number of reactive components (inductors and capacitors).

It should be pointed out that although any ripple height is possible, all the ripples for a particular Chebyshev filter have the same height (a Chebyshev filter is sometimes called an **equiripple filter**) but are not equally spaced horizontally (see Figure 3.10, showing a fifth-order Chebyshev filter). The amount of the ripple height is determined by the value of the components and could be obtained by plotting the $A(\omega)$. The gain in the stopband approaches a value that is $6(n - 1)$ dB lower than the comparable Butterworth filter, where n is the order of the filters.

The Butterworth gain formula is given by Eq. 3.7. What is the Chebyshev gain formula? It is quite a bit more complicated and is given by the following formula:

$$A(\omega) = \frac{k}{\sqrt{1 + \varepsilon^2 C_n^2(\omega/\omega_0)}} \tag{3.8}$$

TABLE 3.2 CHEBYSHEV POLYNOMIALS

Order, n	$C_n(\omega)$
0	1
1	ω
2	$2\omega^2 - 1$
3	$4\omega^3 - 3\omega$
4	$8\omega^4 - 8\omega^2 + 1$
5	$16\omega^5 - 20\omega^3 + 5\omega$
6	$32\omega^6 - 48\omega^4 + 18\omega^2 - 1$
7	$64\omega^7 - 112\omega^5 + 56\omega^3 - 7\omega$
8	$128\omega^8 - 256\omega^6 + 160\omega^4 - 32\omega^2 + 1$

where ε is related to the ripple height and $C_n(\omega/\omega_0)$ is called the Chebyshev polynomial.

There is a different Chebyshev polynomial for each order, n. $C_0(\omega) = 1$ and $C_1(\omega) = \omega$. Higher orders of the Chebyshev polynomial can be calculated by the following recursion formula:

$$C_n(\omega) = 2\omega C_{n-1}(\omega) - C_{n-2}(\omega) \tag{3.9}$$

By using this recursion formula, Table 3.2 can be constructed.

A study of Table 3.2 shows that for $\omega = 0$, $C_n(\omega) = 0$ or ± 1 and for $\omega = 1$, $C_n(\omega) = C_n(1) = 1$. It can be shown that if ω lies between 0 and ω_0 (which is the passband), then $C_n(\omega)$ lies between -1 and $+1$ and $C_n^2(\omega)$ lies between 0 and 1.

Some of the Chebyshev polynomials of Table 3.2 have been plotted in Figure 3.11. The curves look like Lissajous patterns on an oscilloscope. In fact, the

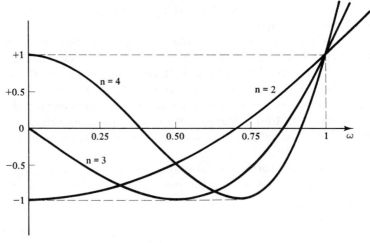

Figure 3.11

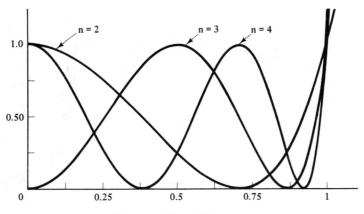

Figure 3.12

mathematical description of Lissajous patterns are Chebyshev polynomials. The number of half-ripples in Figure 3.11 does not seem to agree with the order. But, remember, it is not C_n that is used in Eq. 3.8, it is C_n^2. A plot of C_n^2 is shown in Figure 3.12. Now we can see the correspondence between half-ripples and order. When C_n^2 is inserted into Eq. 3.8, the gain functions plotted in Figure 3.13 result. These are for $\varepsilon = 1$.

Returning to Eq. 3.8, we see that because of the range of C_n, $A(\omega)$ will have a maximum of 1 and a minimum of $1/\sqrt{1 + \varepsilon^2}$ in the passband, as shown in Figure 3.10. The decibel difference is $20 \log 1 - 20 \log (1/\sqrt{1 + \varepsilon^2})$, or just $20 \log (1/\sqrt{1 + \varepsilon^2})$. If decibel is given, ϵ can be calculated as follows:

$$\varepsilon = \sqrt{10^{\mathrm{dB}/10} - 1} \tag{3.10}$$

It can be shown that a minimum occurs at $\omega/\omega_0 = 1$ and a maximum or a minimum occur at $\omega/\omega_0 = 0$. In addition, there can be other minima and maxima between $\omega/\omega_0 = 0$ and 1, depending on the order of the filter.

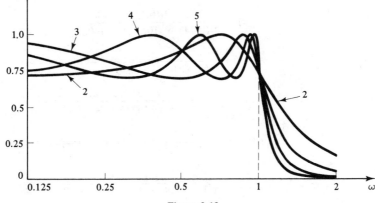

Figure 3.13

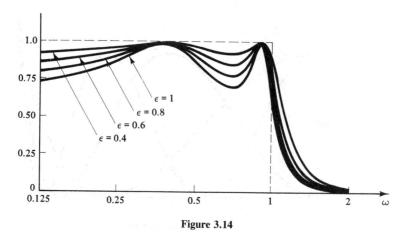

Figure 3.14

The factor, ε, is what determines the weighting of the Chebyshev polynomial. If $\varepsilon = 0$, the curve is a horizontal line and no filtering is done. The effect of varying ε can be seen in Figure 3.14, where the order is all the same ($n = 4$).

Example 3.9

Prove mathematically that the circuit in Figure E3.9 represents a Chebyshev filter.

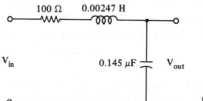

Figure E3.9

Solution   By the voltage-divider rule:

$$G(s) = \frac{1/(0.145 \times 10^{-6} s)}{1/(0.145 \times 10^{-6} s) + 0.00247s + 100}$$

$$= \frac{1}{3.58 \times 10^{-10} s^2 + 1.45 \times 10^{-5} s + 1}$$

$$G(j\omega) = \frac{1}{(1 - 3.58 \times 10^{-10}\omega^2) + j1.45 \times 10^{-5}\omega}$$

$$A(\omega) = \frac{1}{\sqrt{(1 - 3.58 \times 10^{-10}\omega^2)^2 + (1.45 \times 10^{-5}\omega)^2}}$$

$$= \frac{1}{\sqrt{1 - 5.06 \times 10^{-10}\omega^2 + 12.8 \times 10^{-20}\omega^4}}$$

This is a second-order filter (because of the factor ω^4), but is it Chebyshev?   From Eq. 3.8 and Table 3.2, the Chebyshev response is

$$A_{Ch_2}(\omega) = \frac{k}{\sqrt{1 + \epsilon^2(2\omega^2 - 1)^2}} = \frac{k}{\sqrt{(1 + \epsilon^2) - \epsilon^2 4(\omega^2/\omega_0^2) + \epsilon^2 4(\omega^4/\omega_0^4)}}$$

If our equation can be made to fit this one, it is Chebyshev.   In one case, the ratio of the third term to the second is $-2.53 \times 10^{-10}\omega^2$, and in the other case it is $-\omega^2/\omega_0^2$. These ratios are equal if $\omega_0 = 62{,}800$ rad/s or 10 kHz.   Then, comparing the ratio of third terms to first terms, we have

$$\frac{4\epsilon^2\omega^4}{62{,}800^4(1 + \epsilon^2)} = 12.8 \times 10^{-20}\omega^4 \qquad \text{or} \qquad \epsilon = 1$$

Thus any second-order circuit of this type is Chebyshev, if not Butterworth, since an ϵ and a ω_0 could have been found, regardless, as long as center term is negative.

3.7 INVERSE CHEBYSHEV FILTER

Occasionally, it is desired to have a filter that drops quickly after leaving the passband, like the Chebyshev, but without the ripples in the passband.   In that case we can use a filter called the **inverse Chebyshev filter**.   In this case the price we pay is ripples in the stop-band.   This is shown in Figure 3.15.   It can be shown that stopband ripples of equal height will be produced if the Chebyshev polynomials of Table 3.2 are used in the following defining equation:

$$A(\omega) = \frac{k\epsilon C_n(\omega_0/\omega)}{\sqrt{1 + \epsilon^2 C_n^2(\omega_0/\omega)}} \tag{3.11}$$

where ω_0 is not the cutoff frequency but is the frequency at the beginning of the stopband.

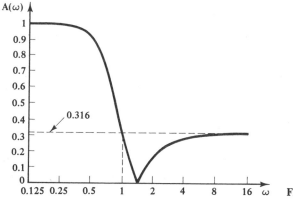

Figure 3.15

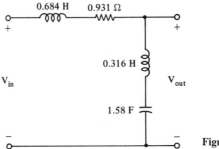

Figure 3.16

Figure 3.16 shows a second-order inverse Chebyshev filter (note: three re-active components), whose stopband ripples lie between a gain of 0.316 and zero (i.e., between -10 dB and $-\infty$ dB). The ε is 0.333 and the ω_0 is 1. The stopband gain is given by $\varepsilon/\sqrt{1 + \epsilon^2}$. Notice that in the stopband, the minimums in the ripple band are all zero gain. In general, there is no advantage to be gained by using the inverse Chebyshev instead of the Chebyshev. If one meets a given specification, the other will also. We will not go into any further detail in this book concerning inverse Chebyshev filters.

Example 3.10

Show that the circuit shown in Figure 3.16 is an inverse Chebyshev filter.

Solution   Using the voltage-divider rule, the transfer function is

$$G(s) = \frac{0.316s + 1/(1.58s)}{s + 0.931 + 1/(1.58s)} = \frac{0.316 + 0.633/s^2}{1 + 0.931/s + 0.633/s^2}$$

If $j\omega$ is substituted for s, then

$$A(\omega) = \frac{0.316 - 0.633/\omega^2}{\sqrt{(1 - 0.633/\omega^2)^2 + 0.866/\omega^2}} = \frac{0.316 - 0.633/\omega^2}{\sqrt{1 - 0.4/\omega^2 + 0.4/\omega^4}}$$

Since the numerator contains the square of ω, this is perhaps a second-order inverse Chebyshev filter. Substituting $C_2(\omega)$ from Table 3.2 into Eq. 3.11 yields

$$A_{\text{IC}}(\omega) = \frac{-\varepsilon + 2\varepsilon(\omega_0/\omega)^2}{\sqrt{1 + \varepsilon^2(2(\omega_0/\omega)^2 - 1)^2}} = \frac{-\varepsilon + 2\varepsilon(\omega_0/\omega)^2}{\sqrt{1 + \varepsilon^2 - 4\varepsilon^2(\omega_0/\omega)^2 + 4\varepsilon^2(\omega_0/\omega)^4}}$$

$$= \frac{-(\varepsilon/\sqrt{1 + \varepsilon^2}) + 2(\varepsilon/\sqrt{1 + \varepsilon^2})(\omega_0/\omega)^2}{\sqrt{1 - 4[\varepsilon/(1 + \varepsilon^2)](\omega_0/\omega)^2 + 4[\varepsilon^2/(1 + \varepsilon^2)](\omega_0/\omega)^4}}$$

This is identical to the equation above for $A(\omega)$ if $\omega_0 = 1$ and $\varepsilon = 0.333$ (or $\varepsilon/\sqrt{1 + \varepsilon^2} = 0.316$). Thus this is an inverse Chebyshev low-pass filter and $A(\infty) = 0.316$ (see Figure 3.15).

3.8 ELLIPTIC (CAUER) FILTER

The third type of non-Butterworth filter is the elliptic filter, so-called because mathematically it involves Jacobi elliptic sine functions in its derivation. This is sometimes called a **Cauer filter**, after the person who first proposed such a filter. This filter has the property that it contains an equiripple band in the stopband as well as in the passband, thus combining the features of the Chebyshev and inverse Chebyshev into one filter (see Figure 3.17). The defining equation is

$$A(\omega) = \frac{k}{\sqrt{1 + \varepsilon^2 R_n^2(\omega)}} \tag{3.12}$$

This looks similar to the Chebyshev equation (Eq. 3.8), except that $R_n(\omega)$ is not a polynomial. Instead, $R_n(\omega)$ is a ratio of two polynomials as follows:

$$R_n(\omega) = \prod_{i=1}^{n/2} \frac{\omega_{2i-1}^2 - \omega^2}{1 - \omega_{2i-1}^2 \omega^2} \qquad \text{(for n even)} \tag{3.13}$$

$$\omega \prod_{i=1}^{\frac{n-1}{2}} \frac{\omega_{2i}^2 - \omega^2}{1 - \omega_{2i}^2 \omega^2} \qquad \text{(for n odd)} \tag{3.14}$$

where Π symbolizes the product of a series of terms, n is the order of the filter, and ω_{2i-1} and ω_{2i} are the frequencies (in rad/s) at which the maximum gain occurs (or the reciprocal of the frequency at which the gain is zero).

If it is required to analyze a circuit to see if it is an elliptic filter, this can be done by a method similar to that used in Example 3.10, by comparing the gain equation for the actual circuit with Eq. 3.13 or 3.14; however, this can be quite a

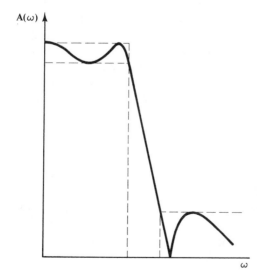

Figure 3.17

tedious process. Another method might be to plot the resulting gain equation, or better yet, just enter the circuit components into the NETWORK program and see if the plot looks like an elliptic filter.

Only the Butterworth and the Chebyshev filters have poles but no zeros. Of these, the Chebyshev drops from the passband to the stopband most quickly (the shortest transition band). If we consider the "all-pole" filters and the filters that have poles and zeros (the inverse Chebyshev and the elliptic filters), it is seen that the elliptic filter has the swiftest transition and thus would make the best filter for a given order, *n*.

For those filters that produce ripples, the order of the filter is usually equal to the number of half-ripples, but this is not true in all cases. This is explored more in Chapter 4.

3.9 BESSEL FILTER

The low-pass filters considered up until now have been rated as to how well their gain response approximates the brick-wall ideal (see Figure 3.7). Very little has been said about the phase response (see Figure 3.5b). If the gain response is almost uniform in the passband of a low-pass Butterworth filter and many low frequencies are present at the input simultaneously, one might think that the output would be a faithful replica of the input. Yet the output is distorted because some frequencies are shifted in phase. In fact, an additional requirement for undistorted output is that there be no phase shift or that each frequency be shifted in phase by an amount proportional to the frequency.

Thus if a 100-Hz signal is shifted 36°, the time shift is

$$T_P = \frac{1}{f}\frac{\text{shift}}{360°} = \frac{1}{100}\frac{36°}{360°} = 1 \text{ ms} \qquad (3.15)$$

If a 200-Hz signal is shifted a proportionate amount of degrees, namely 72°, formula (3.15) will also yield 1 ms. If all the frequencies are shifted an amount proportional to their frequency, all frequencies will be shifted by the same amount of time. This means that the output is merely a time-shifted, but otherwise undistorted version of the input.

A filter that approaches this ideal constant time delay (at least in the passband) is the Bessel filter. If the time delay in the stopband is not constant, this does not matter, since those frequencies are practically filtered out anyway. Although the Bessel filter has nearly constant time delay, its gain response is not nearly as good as the Butterworth filters.

Figure 3.18 compares the time delay of a second-order Bessel filter to that of a second-order Butterworth filter, both being normalized filters. Figure 3.19 compares the gain response of these two filters. Although the Bessel filter appears to drop in the stopband more slowly, this is perhaps an unfair comparison, since the two filters do not reach cutoff at the same frequency. By frequency shifting

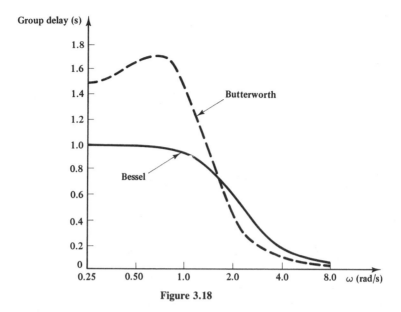

Figure 3.18

the Bessel filter so that its 3-dB cutoff occurs at 1 rad/s also (to be explained in a later chapter), a fair comparison can be made (see Figure 3.20).

Therefore, the Bessel filter finds application where little or no distortion of a multifrequency signal can be tolerated in the passband, but a more gradual drop-off to the stopband can be tolerated.   Since the human ear will not notice phase distortion in voice communication, a Bessel filter is not necessary.   However, in

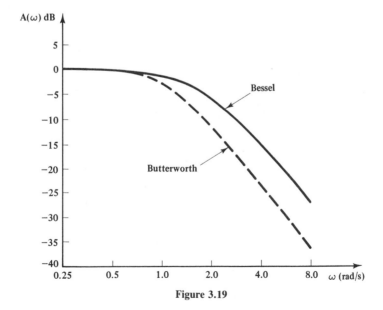

Figure 3.19

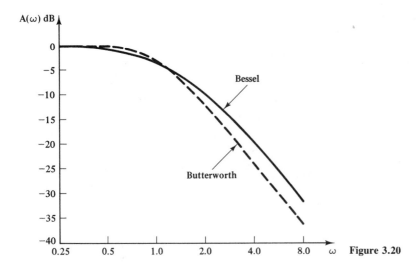

Figure 3.20

a video signal, phase distortion can cause quite noticeable effects on the picture quality; so this might be an area where Bessel filters could help.

The transfer function for a Bessel filter is

$$G(s) = \frac{k}{B_n(s)} \tag{3.16}$$

where k is an arbitrary constant, $B_n(s)$ is called a Bessel polynomial, and n is the order of the filter. The formula for Bessel polynomials is

$$B_n(s) = \sum_{k=0}^{n} \frac{(2n - k)! \, s^k}{2^{n-k}k! \, (n - k)!} \tag{3.17}$$

Once the Bessel polynomials for two successive orders have been found, any succeeding polynomials can be found by the following recursive formula:

$$B_n(s) = (2n - 1)B_{n-1}(s) + s^2 B_{n-2}(s) \tag{3.18}$$

The first six Bessel polynomials are listed in Table 3.3.

TABLE 3.3   BESSEL POLYNOMIALS

Order, n	$B(s)$
1	$s + 1$
2	$s^2 + 3s + 3$
3	$s^3 + 6s^2 + 15s + 15$
4	$s^4 + 10s^3 + 45s^2 + 105s + 105$
5	$s^5 + 15s^4 + 105s^3 + 420s^2 + 945s + 945$
6	$s^6 + 21s^5 + 205s^4 + 1260s^3 + 4725s^2 + 10{,}395s + 10{,}395$

Example 3.11

Find $B_4(s)$, using $B_3(s)$ and $B_2(s)$.

Solution

$$B_4(s) = (2 \times 4 - 1)(s^3 + 6s^2 + 15s + 15) + s^2(s^2 + 3s + 3)$$

$$= s^4 + 10s^3 + 45s^2 + 105s + 105 \quad \text{(agrees with table)}.$$

The gain function, $A(\omega)$, can be found from the transfer function, $G(s)$, as was done in Section 3.3. At that time, the phase function, $\alpha(\omega)$ was also found (see Eq. 3.5). Now we define a new concept called the **group time delay**, $T(\omega)$, as follows:

$$T(\omega) = -\frac{d}{d\omega}\alpha(\omega) \qquad \text{seconds} \qquad (3.19)$$

$T(\omega)$ is called the group time delay because it is the rate of change of phase with respect to frequency for small changes in frequency (over a small "group" of frequencies). Although $\alpha(\omega)$ is a transcendental function (see Eq. 3.6), its derivative is a fraction involving polynomials in ω, and thus is easier to handle (see Example 3.12). The group delay for normalized second order filters is shown in Figure 3.18. Bessel filters are always low-pass filters and the delay function is maximally flat at $\omega = 0$. This means that not only the first derivative of the time delay, but all higher derivatives of the time delay, will be zero at $\omega = 0$. Another property of Bessel filters is that the time delay for normalized Bessel filters of any order is 1 second at ω near zero (see Figure 3.18). However, if the frequency is scaled upward by a factor k_f, so as to get a higher cutoff frequency, the time delay is divided by k_f.

Example 3.12

Find the mathematical expression for the Bessel filter plotted in Figure 3.18.

Solution   Since it was stated previously that this was a second-order filter, using Table 3.3 we get

$$G(s) = \frac{1}{s^2 + 3s + 3}$$

Therefore,

$$G(j\omega) = \frac{1}{(3 - \omega^2) + j3\omega}$$

Further,

$$\alpha(\omega) = -\tan^{-1}\frac{3\omega}{3 - \omega^2}$$

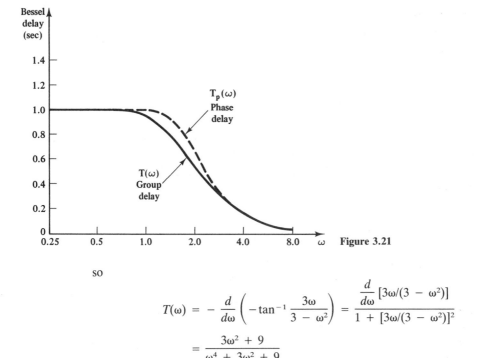

Figure 3.21

so

$$T(\omega) = -\frac{d}{d\omega}\left(-\tan^{-1}\frac{3\omega}{3-\omega^2}\right) = \frac{\frac{d}{d\omega}[3\omega/(3-\omega^2)]}{1+[3\omega/(3-\omega^2)]^2}$$

$$= \frac{3\omega^2 + 9}{\omega^4 + 3\omega^2 + 9}$$

For instance, $T(0) = 1$ second and $T(1) = 0.923$ second.

As was pointed out earlier, the group time delay, $T(\omega)$, is the most common and easiest concept to work with. However, the actual time delay of any particular frequency is called $T_P(\omega)$, the phase delay time, and is given by Eq. 3.14, which in terms of radians is

$$T_P(\omega) = -\frac{\alpha(\omega)}{\omega} \tag{3.20}$$

For instance, for the filter of Example 3.12, it would be

$$T_P(\omega) = \frac{1}{\omega}\tan^{-1}\frac{3\omega}{3-\omega^2}$$

When this is calculated and plotted in Figure 3.21, and compared with $T(\omega)$, the difference in the two concepts is seen.

SUMMARY

Filters are devices that selectively let signals of certain frequencies go through the device with little change in magnitude or phase, while greatly changing the magnitude or phase of signals of other frequencies. In this chapter you saw that many technical fields employ filters, not just the electronics field. Also, filters can employ passive or active circuits, and can be analog or digital in nature, although the digital filter is not introduced until the final chapter.

Circuits, such as discussed in Chapter 2, are actually filters if their frequency response has a definite purpose. The purpose may be to pass low frequencies only, high frequencies only, and so on.

In this chapter we have dealt mainly with passive filters. It describes some typical shapes, such as Butterworth, Chebyshev, elliptic, and Bessel, and shows how to find or recognize the gain equation of each. This equation can then be plotted point by point as a Bode plot to confirm its shape. If the circuit is given, an alternative method is to enter the components into a computer program, called NETWORK, which will do all the calculations and the Bode plotting as well.

PROBLEMS

Sections 3.1 and 3.2

3.1. Based on your knowledge of how inductors and capacitors behave at low and high frequencies, how would you describe the filters shown in Figure P3.1?

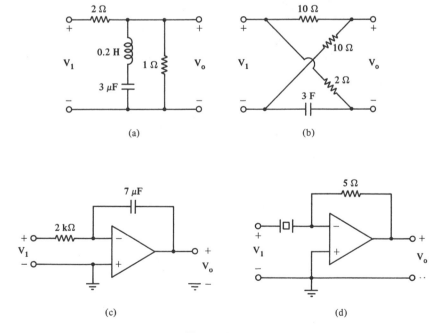

Figure P3.1

3.2. Find the expression for $G(s) = V_o(s)/V_1(s)$ for the circuit shown in Figure P3.2. What type of filter is it?

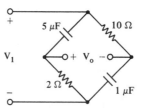

Figure P3.2

3.3. In Figure P3.3, z^{-1} means that the output of that block equals the input but is delayed by one sampling period (1 ms). The -1 is an inverter and the $\rightarrow\oplus\rightarrow$ is an adder. Develop a computer program to test this circuit at .25, 2.5, 25 and 250 Hz using a sampling period of 1 ms. What are the peak amplitudes at these frequencies and what type filter is it?

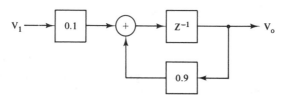

Figure P3.3

3.4. Consider the frequency response shown in Figure P3.4.
   (a) Is it Butterworth, Chebyshev, inverse Chebyshev, elliptic, or none of these?
   (b) What is the passband ripple (in decibels)?
   (c) What is the decibel-level specification in the stopband?
   (d) What is the passband frequency, stopband frequency, and order?

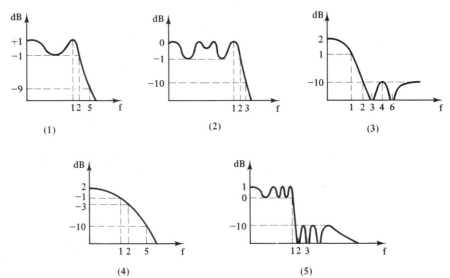

Figure P3.4

Sections 3.3, 3.4 and 3.5

3.5. For Figure P3.5, $G(s) = V_o/V_1$.
   (a) Find $G(s)$.
   (b) Find $|G(j\omega)|$.
   (c) Find ω_0, where asymptotes meet.
   (d) Find $A(\omega_0)$ in decibels.
   (e) Is it or isn't it Butterworth?   State your reasons.

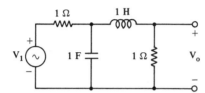

Figure P3.5

3.6. Repeat Problem 3.5 for Figure P3.6.

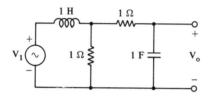

Figure P3.6

3.7. By using the NETWORK computer program, get a plot of the circuit shown in Figure P3.7, and determine if it is Butterworth, Chebyshev, and so on.   Also, find its 3-dB cutoff frequency (if it is Butterworth) or the frequency and decibel levels corresponding to the ripple bands if it is not Butterworth.

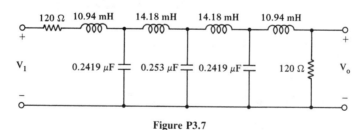

Figure P3.7

3.8. Repeat Problem 3.7 for Figure P3.8.

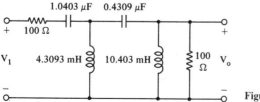

Figure P3.8

3.9. Repeat Problem 3.7 for Figure P3.9.

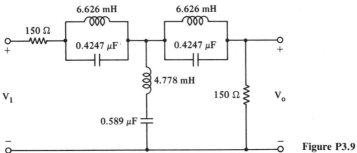

Figure P3.9

3.10. Repeat Problem 3.7 for Figure P3.10.

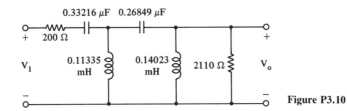

Figure P3.10

3.11. Repeat Problem 3.7 for Figure P3.11.

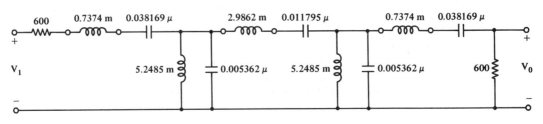

Figure P3.11

3.12. Repeat Problem 3.7 for Figure P3.12.

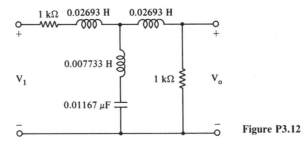

Figure P3.12

Sections 3.6, 3.7 and 3.8

3.13. Does the following gain magnitude function represent a Chebyshev, inverse Chebyshev, or elliptic low-pass filter?

$$A(\omega) = \frac{1}{\sqrt{\omega^6 - 6\omega^4 + 4\omega^2 + 4}}$$

3.14. Repeat Problem 3.13 if

$$A(\omega) = \frac{2}{\sqrt{\omega^6 - 6\omega^4 + 9\omega^2 + 4}}$$

3.15. Repeat Problem 3.13 if

$$A(\omega) = \frac{3\omega^2 - 4}{\sqrt{\omega^6 + 9\omega^4 - 24\omega^2 + 16}}$$

3.16. Repeat Problem 3.13 if

$$A(\omega) = \frac{6\omega^2 - 8}{\sqrt{\omega^6 - 6\omega^4 + 9\omega^2 + 4}}$$

3.17. The filter shown in Figure P3.17 is a Chebyshev filter and $G(s) = V_0/V_1$.
  (a) Find $G(s)$.
  (b) Is it LP or HP?
  (c) Plot the asymptotes of $G(s)$ from 0.1 to 10 rad/s.
  (d) What are the roll-off rates of each asymptote?
  (e) From $G(s)$, find $A(\omega)$.
  (f) Calculate $A(\omega)$ for the following 27 values of ω and convert to decibels and plot the 27 points on the graph from part (c). The frequencies in rad/s are:

0.10	0.50	0.70	0.75	0.80	0.82	0.84	0.86	0.88
0.90	0.92	0.94	0.96	0.98	1.00	1.02	1.04	1.06
1.08	1.10	1.20	1.50	2.00	3.00	5.00	7.00	10.00

  (g) Does the plot from part (f) agree with that from part (c) at the extreme frequencies?
  (h) Any plot has its maximum and minimum where its slope is zero. Differentiate $A(\omega)$, set it equal to zero, and solve for ω_{max}.
  (i) Substitute ω_{max} into $A(\omega)$ to find the maximum gain. Convert it to decibels.
  (j) Why isn't the plot from part (f) 3 dB away from the asymptote intersection at the breakpoint "as it is supposed to be"?

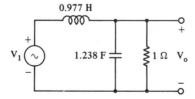

0.977 H

V_1 1.238 F $1\ \Omega$ V_0

Figure P3.17

(k) How much ripple is in the passband?

(l) Find $Z_{in}(s)$ and $|Z_{in}(j\omega)|$.

(m) At what frequency is $|Z_{in}(j\omega)|$ a minimum, and what is the value of $|Z_{in}(j\omega)|$ at this frequency?

(n) Find $Z_{out}(s)$ and $|Z_{out}(j\omega)|$.

(o) At what frequency is $|Z_{out}(j\omega)|$ a maximum, and what is the value of $|Z_{out}(j\omega)|$ at this frequency?

Section 3.9

3.18. From Figure P3.18, which is a Bessel filter?

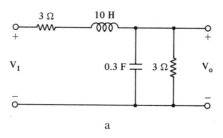

a

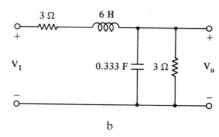

b   **Figure P3.18**

3.19. Which of the following four transfer functions represents a low-pass Bessel filter?

(a) $\dfrac{s + 1.356}{s^2 + 3.678s + 6.458} - \dfrac{1}{s + 2.322}$ **(b)** $\dfrac{1}{s^2 + 3s + 1}$

(c) $\dfrac{1}{s^2(3s + 18) + 45(s + 1)}$ **(d)** $\dfrac{1}{3s^2 + 3s + 1}$

4

synthesis of passive filters

OBJECTIVES

After completing this chapter, you should be able to:

- Construct specification walls on a "gain versus frequency" plot, given a verbal description of the specifications.
- Interpret a plot of gain boundaries and identify A_{PB}, A_Ω, and Ω, and use the proper nomograph to establish the order.
- Select the proper circuit topology to suit the given specs.
- Use the tables properly to find the prototype component values.
- Perform frequency and impedance scaling on the prototype values to get the final values of all components.
- Find the proper components and their values so as to convert a prototype LP filter to HP, BP, or BR filters.
- Use the computer program to check a designed circuit to make sure that it meets the original specification.

INTRODUCTION

Technologists are sometimes called upon to analyze a circuit and sometimes to synthesize a circuit. The latter is usually more difficult and can lead to an infinite number of answers. For instance, to find the total resistance of a series network consisting of a 5-Ω and a 10-Ω resistor, the total resistance is just 15 Ω. This is called analysis. However, if we are required to build a two-resistor network that has a total resistance of 15 Ω, there are an infinite number of circuits that would do the job: series: 5 and 10; 4 and 11; 3 and 12; and so on; and parallel: 20 and 60; 24 and 40; and so on. This is called **synthesis** and is almost an art in itself.

4.1 FILTER SPECIFICATION BY BODE PLOT

The beginning of filter synthesis or design is either to have a frequency response (Bode plot) or transfer function, $G(s)$. Most of this chapter is concerned with the design of passive low-pass filters. Once we master this, the design of the other types becomes much easier; in fact, the technique used for these other types is first to design an appropriate low-pass filter and then use a conversion process to get the required high-pass, bandpass, and so on, from this low-pass "prototype."

So let us begin by designing a LP filter whose Bode plot is as shown in Figure 4.1. From our knowledge gained in Section 2.13, we see that this plot has two asymptotes, one horizontal and one dropping at a rate of 20 dB/decade. The breakpoint is where these two asymptotes meet, namely at ω = 2 rad/s. This means that the "cutoff" frequency is 2 rad/s for this filter. This requires a factor of $(s + 2)$ in the denominator and no such factors in the numerator. Thus $G(s)$ = $1/(s + 2)$. If $G(s)$ is defined as $V_o(s)/V_1(s)$, what circuit will yield this $G(s)$? This is where the situation gets tricky. Let us just state that either of the two circuits shown in Figure 4.2 will do the job. The student is urged to confirm this, by applying the voltage-divider rule to these circuits in order to find $G(s)$.

There are probably many other circuits that would work. Of course, if someone had given us the circuit but not the values, you probably could come up with the component values after a little thought. The art of coming up with the

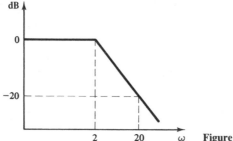

Figure 4.1

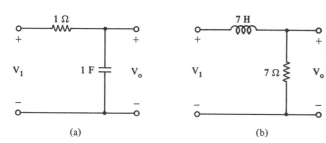

Figure 4.2

proper circuit topology and values is called network synthesis, and entire books have been written on the subject.

For instance, if $G(s) = (s^2 + 2)/(s^3 + 4s^2 + 6s + 12)$, then it might be difficult to "guess" the correct network topology and just as difficult to find the element values. We will not go into the subject of synthesis that deeply here. However, if we are interested in low-pass filters specifically, there are a few standard topologies that will be sufficient for all our work. These circuits and the element values are given later in tables.

The filter specification is not usually given as a precise frequency plot, but rather as a bounded region into which the final design must fit. These bounds usually consist of vertical and horizontal lines on Bode graph paper, as shown in Figure 4.3. Translating the specification of Figure 4.3 into words, we have a low-pass filter that has a passband extending to 10 kHz with a variation not to exceed 3 dB, a stopband starting at 20 kHz, during which the response must be at least 10 dB down from the peak passband value. The area between 10 and 20 kHz is the transition band. As you can see, this description is quite lengthy. Figure 4.3 says the same thing, but much more clearly and concisely. Figure 4.4 shows Bode plots of two filters and the specification walls. It is seen that neither of these two filters fits the specification, so they are unacceptable.

To meet the specification of Figure 4.3, we require a filter that drops much faster than the two filters of Figure 4.4. If we had a slope of -40 dB/decade (or -12 dB/octave), we could meet this specification. This would require more reactive components, that is, the order of the filter would have to be greater.

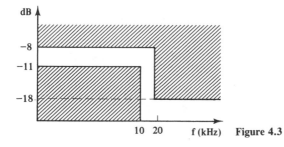

Figure 4.3

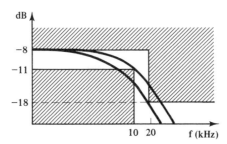

Figure 4.4

4.2 USING NOMOGRAPHS TO DETERMINE FILTER ORDER

There are many filter forms (Butterworth, Chebyshev, etc.) that would satisfy Figure 4.3, and although a first-order filter will not work, a second-, third-, fourth-, or *n*th-order filter would. Which form and which order shall we choose? Unless there are other restrictions given in the problem, it is usually desirable to choose the lowest-order filter that will do the job; or, perhaps the filter with the least number of components (this usually means the lowest order, also). But sometimes this means that an elliptic or Chebyshev filter of a lower order will do the same job as a Butterworth of higher order. If the same order is required for each form, a Butterworth filter is usually chosen.

We could try several filters and make Bode plots of each one, then find which ones fell into the specification walls, and then pick the simplest of these as our choice. But it would be nice if we could determine the required filter order before actually starting to design the filter. This is possible if we use the nomographs shown in Figures 4.5 to 4.7. Four quantities are involved in their use: A_{PB}(dB), A_{Ω}(dB), Ω, and *n*. They are defined as follows:

A_{PB}(dB) = allowable passband variation, dB

A_{Ω}(dB) = difference between the maximum dB level in the passband and the maximum dB level in the stopband

Ω = ratio of stopband frequency to passband frequency

n = order of the filter (must be a whole number)

The procedure is as follows (see Figure 4.8):

1. Draw a line from the A_{PB}(dB) to A_{Ω}(dB) and continue this line until it meets the ordinate of the graph.
2. From this point, move horizontally across the graph until it meets the vertical line drawn upward from Ω.
3. This point may be on an *n* line. If so, this is the order of the filter. If not, the required order is the higher of the two lines between which it lies.

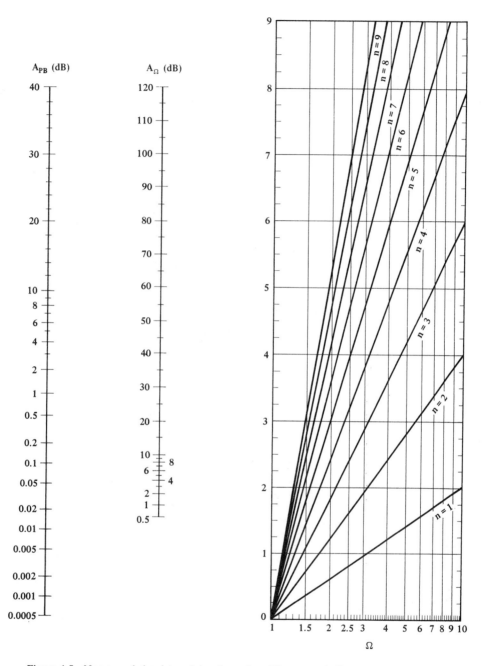

Figure 4.5   Nomograph for determining the order of Butterworth filters.
Reprinted by permission from Huelsman and Allen, "Theory and Design of Active Filters,"
McGraw-Hill Book Company, Inc., 1980.

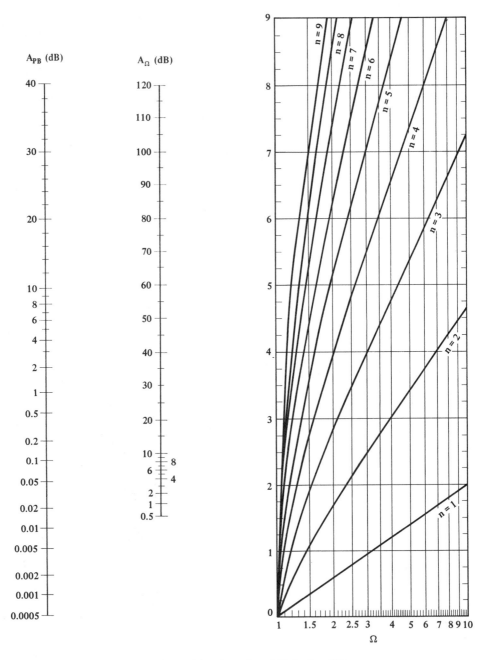

Figure 4.6 Nomograph for determining the order of Chebyshev filters.
Reprinted by permission from Huelsman and Allen, "Theory and Design of Active Filters,"
McGraw-Hill Book Company, Inc., 1980.

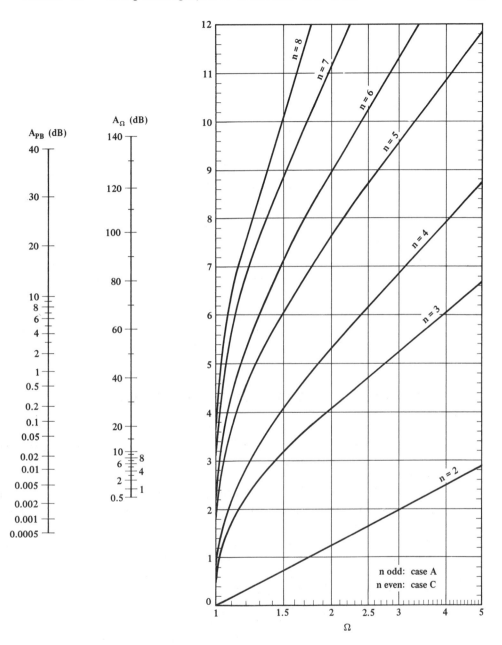

Figure 4.7   Nomograph for determining the order of elliptic filters.
Reprinted by permission from Huelsman and Allen, "Theory and Design of Active Filters,"
McGraw-Hill Book Company, Inc., 1980.

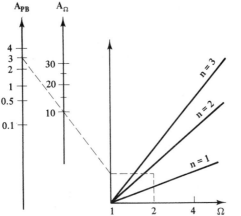

Figure 4.8

Example 4.1

Find the minimum-order Butterworth filter that will satisfy the specs shown in Figure 4.3.

Solution   From Figure 4.3, A_{PB}(dB) = 3, A_{Ω}(dB) = 10, and Ω = 20 kHz/10 kHz = 2. Using the procedure shown in Figure 4.8, we get an n value between 1 and 2, so we must pick n = 2 (a second-order filter).

4.3 USING TABLES TO DESIGN THE LOW-PASS PROTOTYPE FILTER

Now that we have found the order of the required filter, we must find the transfer function, $G(s)$. From $G(s)$, the proper circuit layout (topology) and the element values can be found. However, for a passive circuit, many different topologies are possible. We will settle on a lossless ladder network, called a **Darlington realization.** It is called lossless because no power is consumed in the filter, since all components are inductors and capacitors, except for a resistor at the end, which can represent the load, and in some cases a resistor at the beginning which can represent the source impedance. From the required $A(\omega)$ (see Eqs. 3.7, 3.8, 3.11, 3.12), the $G(s)$ can be found and the circuit topologies and element values can be found. However, the techniques involved are beyond the scope of this book (see the book by Karni listed in the Bibliography).

Instead, we use standard circuits whose element values are read from tables or can be found from computer programs. However, it would be impractical to have tables of values for every type of filter (LP, HP, BP, etc.) and for every cutoff frequency that might be needed. Instead, tables will be used that are good only for LP filters with a cutoff of 1 rad/s. This is called a "prototype" filter. We will see later how we can convert the results to give a filter that will meet our actual specifications.

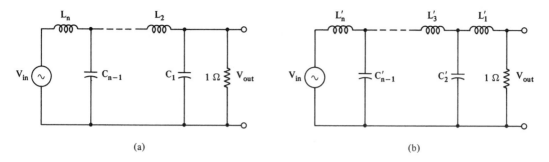

Figure 4.9
Reprinted by permission from Huelsman and Allen, "Theory and Design of Active Filters," McGraw-Hill Book Company, Inc., 1980.

For the Butterworth filter, cutoff can be defined at any arbitrary decibel level. When using the tables, we know that the gain at 1 rad/s will be 3 dB less than the gain at $\omega = 0$: this is stated in the sections of Tables 4.1 to 4.3, which apply to Butterworth filters. From these tables, reference is made to Figures 4.9 to 4.12. Each table refers to at least two circuits that can use the element values of the table. One circuit has an inductor as the first reactive element of the ladder; the other has a capacitor as the first reactive element. All circuits have a load resistor of 1 Ω. Some circuits have a resistor of some value in series with the source; others do not. All circuits assume that the source is a voltage source (except for two of the circuits). After picking the desired circuit topology, it is important that, when using the values, they are placed with the symbols shown closest to the referenced figure in the tables. The following examples illustrate these points:

Example 4.2

Design a Butterworth third-order LP filter, having a 3-dB variation between 0 and 1 rad/s.

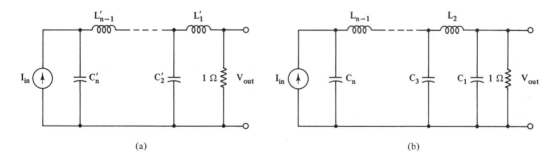

Figure 4.10
Reprinted by permission from Huelsman and Allen, "Theory and Design of Active Filters," McGraw-Hill Book Company, Inc., 1980.

TABLE 4.1 LOW-PASS LOSSLESS LADDER (DARLINGTON) REALIZATION TERMINATED WITH A RESISTOR AT ONE END ONLY

For n even, refer to Figure 4.9a; for n odd, refer to Figure 4.10b.

n	C_1	L_2	C_3	L_4	C_5	L_6	C_7	L_8	C_9	L_{10}
						Butterworth				
						(BW = 1 rad/s at 3 dB cutoff)				
2	0.7071	1.4142								
3	0.5000	1.3333	1.5000							
4	0.3827	1.0824	1.5772	1.5307						
5	0.3090	0.8944	1.3820	1.6944	1.5451					
6	0.2588	0.7579	1.2016	1.5529	1.7593	1.5529				
7	0.2225	0.6560	1.0550	1.3972	1.6588	1.7988	1.5576			
8	0.1951	0.5776	0.9370	1.2588	1.5283	1.7287	1.8246	1.5607		
9	0.1736	0.5155	0.8414	1.1408	1.4037	1.6202	1.7772	1.8424	1.5628	
10	0.1564	0.4654	0.7626	1.0406	1.2921	1.5100	1.6869	1.8121	1.8552	1.5643
						$\frac{1}{2}$-dB ripple Chebyshev				
						(BW = 1 rad/s)				
2	0.7014	0.9403								
3	0.7981	1.3001	1.3465							
4	0.8352	1.3916	1.7279	1.3138						
5	0.8529	1.4291	1.8142	1.6426	1.5388					
6	0.8627	1.4483	1.8494	1.7101	1.9018	1.4042				
7	0.8686	1.4596	1.8675	1.7371	1.9712	1.7254	1.5982			
8	0.8725	1.4666	1.8750	1.7508	1.9980	1.7838	1.9571	1.4379		
9	0.8752	1.4714	1.8856	1.7591	2.0016	1.8055	2.0203	1.7571	1.6238	
10	0.8771	1.4748	1.8905	1.7645	2.0197	1.8165	2.0432	1.8119	1.9816	1.4539

n	L_1'	C_2'	L_3'	C_4'	L_5'	C_6'	L_7'	C_8'	L_9'	C_{10}'
						1-dB ripple Chebyshev				
						(BW = 1 rad/s)				
2	0.9110	0.9957								
3	1.0118	1.3332	1.5088							
4	1.0495	1.4126	1.9093	1.2817						
5	1.0674	1.4441	1.9938	1.5908	1.6652					
6	1.0773	1.4601	2.0270	1.6507	2.0491	1.3457				
7	1.0832	1.4694	2.0437	1.6736	2.1192	1.6489	1.7118			
8	1.0872	1.4751	2.0537	1.6850	2.1453	1.7021	2.0922	1.3691		
9	1.0899	1.4790	2.0601	1.6918	2.1583	1.7213	2.1574	1.6707	1.7317	
10	1.0918	1.4817	2.0645	1.6961	2.1658	1.7306	2.1803	1.7215	2.1111	1.3801

For n odd, refer to Figure 4.9b; for n even, refer to Figure 4.10a.

Reprinted by permission from Weinberg, Louis, "Network Analysis and Synthesis," McGraw-Hill Book Company, Inc., 1962, (reprinted by R. E. Krieger, 1976).

TABLE 4.2 LOW-PASS LOSSLESS LADDER (DARLINGTON) REALIZATION TERMINATED WITH A 1-Ω RESISTOR AT EACH END

For n even, refer to Figure 4.11a; for n odd, refer to Figure 4.12b.

n	C_1	L_2	C_3	L_4	C_5	L_6	C_7	L_8	C_9	L_{10}
						Butterworth				
						(BW = 1 rad/s at 3 dB)				
2	1.4142	1.4142								
3	1.0000	2.0000	1.0000							
4	0.7654	1.8478	1.8478	0.7654						
5	0.6180	1.6180	2.0000	1.6180	0.6180					
6	0.5176	1.4142	1.9319	1.9319	1.4142	0.5176				
7	0.4450	1.2470	1.8019	2.0000	1.8019	1.2470	0.4450			
8	0.3902	1.1111	1.6629	1.9616	1.9616	1.6629	1.1111	0.3902		
9	0.3473	1.0000	1.5321	1.8794	2.0000	1.8794	1.5321	1.0000	0.3473	
10	0.3129	0.9080	1.4142	1.7820	1.9754	1.9754	1.7820	1.4142	0.9080	0.3129
						½-dB ripple Chebyshev				
						(BW = 1 rad/s)				
3	1.5963	1.0967	1.5963							
5	1.7058	1.2296	2.5408	1.2296	1.7058					
7	1.7373	1.2582	2.6383	1.3443	2.6383	1.2582	1.7373			
9	1.7504	1.2690	2.6678	1.3673	2.7239	1.3673	2.6678	1.2690	1.7504	
						1-dB ripple Chebyshev				
						(BW = 1 rad/s)				
3	2.0236	0.9941	2.0236							
5	2.1349	1.0911	3.0009	1.0911	2.1349					
7	2.1666	1.1115	3.0936	1.1735	3.0936	1.1115	2.1666			
9	2.1797	1.1192	3.1214	1.1897	3.1746	1.1897	3.1214	1.1192	2.1797	
n	L'_1	C'_2	L'_3	C'_4	L'_5	C'_6	L'_7	C'_8	L'_9	C'_{10}

For n even, refer to Figure 4.12a; for n odd, refer to Figure 4.11b.

Reprinted by permission from Weinberg, Louis, "Network Analysis and Synthesis," McGraw-Hill Book Company, Inc., 1962, (reprinted by R. E. Krieger, 1976).

TABLE 4.3 LOW-PASS LOSSLESS LADDER (DARLINGTON) REALIZATION TERMINATED WITH DIFFERENT RESISTORS AT EACH END, $X = \frac{1}{2}$

For n even, refer to Figure 4.11a; for n odd, refer to Figure 4.12b.

n	C_1	L_2	C_3	L_4	C_5	L_6	C_7	L_8	C_9	L_{10}
						Butterworth				
						(BW = 1 rad/s at 3 dB)				
2	3.3461	0.4483								
3	3.2612	0.7789	1.1811							
4	3.1868	0.8826	2.4524	0.2175						
5	3.1331	0.9237	3.0510	0.4955	0.6857					
6	3.0938	0.9423	3.3687	0.6542	1.6531	0.1412				
7	3.0640	0.9513	3.5532	0.7512	2.2726	0.3536	0.4799			
8	3.0408	0.9558	3.6678	0.8139	2.6863	0.5003	1.2341	0.1042		
9	3.0223	0.9579	3.7426	0.8565	2.9734	0.6046	1.7846	0.2735	0.3685	
10	3.0072	0.9588	3.7934	0.8864	3.1795	0.6808	2.1943	0.4021	0.9818	0.0825
						$\frac{1}{2}$-dB ripple Chebyshev				
						(BW = 1 rad/s)				
2	1.5132	0.6538								
3	2.9431	0.6503	2.1903							
4	1.8158	1.1328	2.4881	0.7732						
5	3.2228	0.7645	4.1228	0.7116	2.3197					
6	1.8786	1.1884	2.7589	1.2403	2.5976	0.7976				
7	3.3055	0.7899	4.3575	0.8132	4.2419	0.7252	2.3566			
8	1.9012	1.2053	2.8152	1.2864	2.8479	1.2628	2.6310	0.8063		
9	3.3403	0.7995	4.4283	0.8341	4.4546	0.8235	4.2795	0.7304	2.3719	
10	1.9117	1.2127	2.8366	1.2999	2.8964	1.3054	2.8744	1.2714	2.6456	0.8104

n	L'_1	C'_2	L'_3	C'_4	L'_5	C'_6	L'_7	C'_8	L'_9	C'_{10}
						1-dB ripple Chebyshev				
						(BW = 1/rads)				
3	3.4774	0.6153	2.8540							
5	3.7211	0.6949	4.7448	0.6650	2.9936					
7	3.7916	0.7118	4.9425	0.7348	4.8636	0.6757	3.0331			
9	3.8210	0.7182	5.0013	0.7485	5.0412	0.7429	4.9004	0.6797	3.0495	

For n even, refer to Figure 4.12a; for n odd, refer to Figure 4.11b.

Reprinted by permission from Weinberg, Louis, "Network Analysis and Synthesis," McGraw-Hill Book Company, Inc., 1962, (reprinted by R. E. Krieger, 1976).

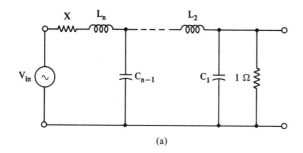

(a)

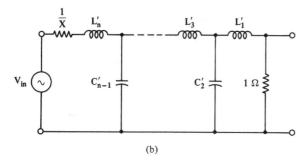

(b)

Figure 4.11
Reprinted by permission from Huelsman and Allen, "Theory and Design of Active Filters," McGraw-Hill Book Company, Inc., 1980.

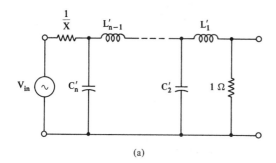

(a)

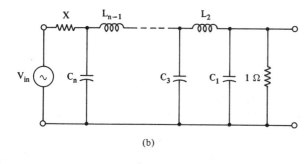

(b)

Figure 4.12
Reprinted by permission from Huelsman and Allen, "Theory and Design of Active Filters," McGraw-Hill Book Company, Inc., 1980.

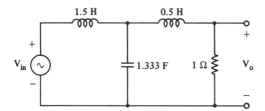

Figure E4.2

Solution   Since any of the three Butterworth tables can be used, let's arbitrarily pick Table 4.1.   At the top of this table it says to use Figure 4.10b if it is an odd-order filter.   At the bottom of this table it says to use Figure 4.9b if it is odd.   Again, arbitrarily pick Figure 4.9b; however, we must associate the table values with the primed symbols shown at the bottom of the table, since the figure note that we are using is at the bottom of the table.   Reading the table, we see that $L_1' = 0.5$, $C_2' = 1.333$, and $L_3' = 1.5$.   Notice that the circuits shown in the figures are expandable to accommodate the higher-order filters by adding as many LC pairs as required (in the dashed area).   Inserting the values into the figure, we get the circuit shown in Figure E4.2.

Example 4.3

Repeat Example 4.2, except with the added requirement that the source be a current source.

Solution   This added requirement restricts our choice to Figure 4.10, the only current-source figure; and since it is an odd order, Figure 4.10b must be used.   According to the note on Table 4.1, the unprimed symbols must now be assigned to the table values.   The final circuit thus looks as shown in Figure E4.3.

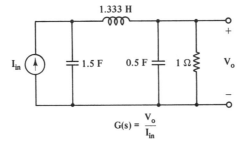

$$G(s) = \frac{V_o}{I_{in}}$$

Figure E4.3

Example 4.4

Realize a Butterworth fourth-order LP filter with a 3-dB cutoff at 0.159 Hz if driven by a voltage source, having a source impedance equal to the load impedance.

Solution   Since 0.159 Hz $= 1$ rad/s and the variation is 3 dB, the Butterworth tables can be used without alteration.   Of the Butterworth figures, only Figures 4.11 and 4.12 have resistors at both ends; and of these, only Figures 4.11a and 4.12a are the even-order case.   Either one can be chosen, by letting $X = 1$ and choosing Table 4.2, the one with the "equal" resistors.   Choosing Figure 4.12a arbitrarily and reading

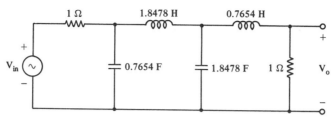

Figure E4.4

the primed symbols, because the note referring to Figure 4.12a is at the bottom of the table, yields the final circuit as shown in Figure E4.4.

Example 4.5

Realize a fifth-order Butterworth filter with a 3-dB cutoff at 1 rad/s if it is to be voltage driven and the source resistance is twice as great as the load resistance.

Solution   This is again a problem using either Figure 4.11 or 4.12, except that the unequal resistor table, Table 4.3, must be used if possible. This table says that $X =$ 0.5. If you look closely at Figures 4.11b and 4.12a, you will see that the source resistor is $1/X$, or $1/0.5$, or 2 Ω, which is exactly twice the load resistor shown on the right end of the circuit. Of course, Figure 4.11b is the one for odd-order filters and requires the primed elements yielding the circuit in Figure E4.5.

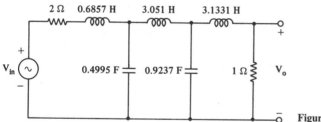

Figure E4.5

In addition to the Butterworth LP prototype filter tables, we will be using Tables 4.1 to 4.3 for 0.5- and 1.0-dB ripple Chebyshev LP prototype filters, Tables 4.4 and 4.5 for 0.1- and 1.0-dB ripple elliptic LP prototype filters, and Table 4.6 for 1.0-second-delay Bessel prototype filters. The next two examples illustrate some of these applications.

Example 4.6

Realize a fourth-order LP Chebyshev filter having 0.5-dB variation in the range from dc to 1 rad/s and driven by a voltage source with a resistance that is half of the load resistance.

Solution   The "unequal" resistor table is Table 4.3, with $X = 0.5$, so we will need to use either Figure 4.11a or 4.12b to get the source resistance $= 0.5$ Ω; but since a fourth-order filter is even, we must choose Figure 4.11a. Notice that if we wanted the source resistance to be one-fourth of the load resistance, no table would be

TABLE 4.4 ELLIPTIC (CAUER) LOW-PASS LOSSLESS LADDER (DARLINGTON) REALIZATION TERMINATED WITH A 1-Ω RESISTOR AT EACH END

Values of elements (henries or farads) in Figure 4.13

n	Ω	A_Ω(dB)	L_1	C_2	L_2	L_3	C_4	L_4	L_5	C_6
3	1.05	1.748	0.3555	0.1537	5.3960	0.3555				
	1.10	3.374	0.4463	0.2699	2.7035	0.4463				
	1.20	6.691	0.5734	0.4498	1.3081	0.5734	0.1-dB passband ripple			
	1.50	14.848	0.7703	0.7456	0.4780	0.7703				
	2.00	24.010	0.8954	0.9376	0.2070	0.8954				
4	1.05	3.284	0.0044	0.1722	4.9376	1.0122	0.8445			
	1.10	6.478	0.1728	0.3276	2.3099	1.0489	0.8942			
	1.20	12.085	0.3714	0.5664	1.0929	1.1194	0.9244			
	1.50	23.736	0.6282	0.9401	0.4073	1.2471	0.9352			
	2.00	36.023	0.7755	1.1765	0.1796	1.3347	0.9338			
5	1.05	13.841	0.7081	0.7663	0.7357	1.1276	0.2014	4.3812	0.0499	
	1.10	20.050	0.8130	0.9242	0.4934	1.2245	0.3719	2.1350	0.2913	
	1.20	28.303	0.9144	1.0652	0.3163	1.3820	0.6013	1.0933	0.5297	
	1.50	43.415	1.0279	1.2152	0.1513	1.6318	0.9353	0.4408	0.8155	
	2.00	58.901	1.0876	1.2932	0.0732	1.7939	1.1433	0.2004	0.9772	
6	1.05	18.727	0.4418	0.7165	0.9091	0.8314	0.3627	2.4468	0.8046	0.9986
	1.10	26.230	0.5763	0.8880	0.6128	0.9730	0.5906	1.3567	0.9431	1.0138
	1.20	36.113	0.7098	1.0627	0.3914	1.1597	0.8741	0.7619	1.0918	1.0246
	1.50	54.202	0.8660	1.2740	0.1855	1.4311	1.2724	0.3301	1.2825	1.0332
	2.00	72.761	0.9513	1.3930	0.0893	1.6013	1.5187	0.1542	1.3952	1.0362
n			C_1'	L_2'	C_2'	C_3'	L_4'	C_4'	C_5'	L_6'

Values of elements (henries or farads) in Figure 4.14.

Reprinted by permission from Huelsman and Allen, "Theory and Design of Active Filters," McGraw-Hill Book Company, Inc., 1980 computed using a program written by David Jose Baezlopez, "Sensitivity and Synthesis of Elliptic Functions," Ph.D. dissertation, University of Arizona, Tucson, 1978.

TABLE 4.5 ELLIPTIC (CAUER) LOW-PASS LOSSLESS LADDER (DARLINGTON) REALIZATION TERMINATED WITH A 1-Ω RESISTOR AT EACH END

Values of elements (henries or farads) in Figure 4.13

n	Ω	A_Ω(dB)	L_1	C_2	L_2	L_3	C_4	L_4	L_5	C_6
3	1.05	8.134	1.0551	0.2522	3.2890	1.0551				
	1.10	11.480	1.2253	0.3747	1.9475	1.2253				
	1.20	16.209	1.4245	0.5254	1.1198	1.4245				
	1.50	25.176	1.6920	0.7334	0.4859	1.6920				
	2.00	34.454	1.8520	0.8590	0.2259	1.8520	1.0-dB passband ripple			
4	1.05	11.322	0.6371	0.3528	2.4104	1.1152	1.3995			
	1.10	15.942	0.8094	0.5404	1.4002	1.1811	1.4500			
	1.20	22.293	1.0033	0.7773	0.7963	1.2662	1.4922			
	1.50	34.179	1.2568	1.1143	0.3436	1.3898	1.5323			
	2.00	46.481	1.4068	1.3237	0.1596	1.4676	1.5507			
5	1.05	24.135	1.5619	0.6756	0.8345	1.5546	0.2658	3.3188	0.8853	
	1.10	30.471	1.6969	0.7751	0.5883	1.7989	0.3992	1.9891	1.1211	
	1.20	38.757	1.8281	0.8701	0.3872	2.0910	0.5635	1.1667	1.3809	
	1.50	53.875	1.9769	0.9769	0.1882	2.4916	0.7936	0.5195	1.7189	
	2.00	69.360	2.0559	1.0339	0.0915	2.7357	0.9356	0.2449	1.9194	
6	1.05	29.133	1.0746	0.8012	0.8130	0.9274	0.5175	1.7150	0.9219	1.6051
	1.10	36.680	1.2206	0.9424	0.5775	1.1090	0.7572	1.0582	1.0168	1.6468
	1.20	46.571	1.3715	1.0863	0.3828	1.3261	1.0511	0.6335	1.1248	1.6850
	1.50	64.661	1.5543	1.2588	0.1878	1.6253	1.4656	0.2866	1.2696	1.7248
	2.00	83.221	1.6566	1.3545	0.0918	1.8086	1.7238	0.1359	1.3573	1.7442
n			C_1'	L_2'	C_2'	C_3'	L_4'	C_4'	C_5'	L_6'

Values of elements (henries or farads) in Figure 4.14.

Reprinted by permission from Huelsman and Allen, "Theory and Design of Active Filters," McGraw-Hill Book Company, Inc., 1980 computed using a program written by David Jose Baezlopez, "Sensitivity and Synthesis of Elliptic Functions," Ph.D. dissertation, University of Arizona, Tucson, 1978.

TABLE 4.6 BESSEL LOW-PASS LOSSLESS LADDER (DARLINGTON) REALIZATION TERMINATED WITH A 1-Ω RESISTOR AT EACH END

For n even, refer to Figure 4.11a; for n odd refer to Figure 4.12b

n	$A(1)$ (dB)	$T(1)$ (sec)	C_1	L_2	C_3	L_4	C_5	L_6	C_7	L_8	C_9
1	−3.0	0.5000	2.0000								
2	−1.6	0.9231	1.5774	0.4226							
3	−0.9	0.9964	1.2550	0.5528	0.1922						
4	−0.6	0.9999	1.0598	0.5116	0.3181	0.1104					
5			0.9303	0.4577	0.3312	0.2090	0.0718				
6			0.8377	0.4116	0.3158	0.2364	0.1480	0.0505			
7			0.7677	0.3744	0.2944	0.2378	0.1778	0.1104	0.0375		
8			0.7125	0.3446	0.2735	0.2297	0.1867	0.1387	0.0855	0.0289	
9			0.6678	0.3203	0.2547	0.2184	0.1859	0.1506	0.1111	0.0682	0.0230
n			L_1'	C_2'	L_3'	C_4'	L_5'	C_6'	L_7'	C_8'	L_9'

Delay = 1 second (at $\omega = 0$)

For n even, refer to Figure 4.12a; for n odd, refer to Figure 4.11b.

Reprinted by permission from Weinberg, Louis, "Network Analysis and Synthesis," McGraw-Hill Book Company, Inc., 1962, (reprinted by R. E. Krieger, 1976).

146

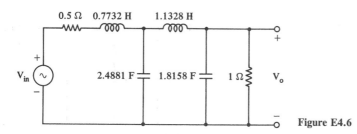

Figure E4.6

available in our book. Of course, it is possible that other filter handbooks could have more extensive tables to cover this situation. Using the unprimed symbols, we get the circuit in Figure E4.6.

Example 4.7

Design a third-order LP elliptic filter having 0.1-dB variation in the passband from dc to 1 rad/s.

Solution Elliptic tables are found in this text only in Tables 4.4 and 4.5, and since we want 0.1-dB ripple, we must use Table 4.4 with Figure 4.13 or 4.14. These are exclusively for elliptic filters. It is completely arbitrary which one we use. The source resistor must equal the load resistor: we have no choice. This means that we cannot build elliptic filters with only one resistor or with unequal resistors, at least not with what we have in this book. As before, each figure can have series-shunt sections added, to get the order required. The right-end termination looks a bit different for the odd orders than for the even orders, as it did in previous figures. Let's pick Figure 4.13 (left part) plus the "odd" termination (right part). However, we see that the left part has too many elements for a third-order filter, so we just do not use the elements marked L_3, L_4, and C_4, when drawing the circuit. After sketching the circuit and substituting the unprimed values from the table, and frequency ratio, $\Omega = 1.5$, we get the circuit shown in Figure E4.7. Omega, Ω, was chosen arbitrarily, since no stopband level or stopband frequency was specified in the problem.

Examples 4.2 to 4.7 have shown filters that meet certain decibel variations in the passband, but we have not specified the actual decibel level. Does a 3-dB filter vary from 3 to 0 dB or from 0 to -3 dB or from -6 to -9 dB? Knowing the general shape of the response and the circuit itself, it is easy to answer this

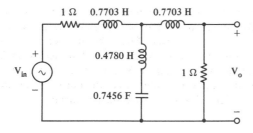

Figure E4.7

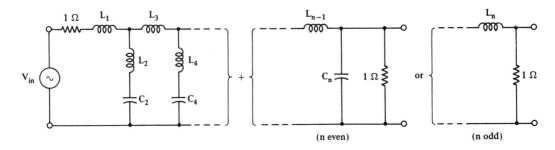

Figure 4.13
Reprinted by permission from Huelsman and Allen, "Theory and Design of Active Filters,"
McGraw-Hill Book Company, Inc., 1980.

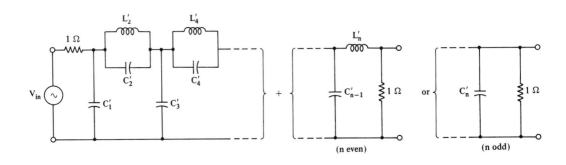

Figure 4.14
Reprinted by permission from Huelsman and Allen, "Theory and Design of Active Filters,"
McGraw-Hill Book Company, Inc., 1980.

question, by just examining $G(0)$. For instance, in Example 4.2, at $\omega = 0$, all
inductors are shorts and all capacitors are opens, so that the output = the input
and $G(0) = 1$, which is $20 \times \log (1) = 0$ dB. Since a Butterworth drops monotonically, the variation is 0 to -3 dB.

Example 4.8

What are the passband levels for Example 4.5?

Solution Shorting the inductors and opening the capacitors leaves us with just two
resistors in series. By the voltage-divider rule (VDR), we get $V_o/V_{in} = 1/(1 + 2)$
$= \frac{1}{3} = -9.5$ dB. Since it is a monotonic filter, the filter starts at -9.5 dB and at 1
rad/s it is -12.5 dB. If we wanted to use this filter, but insisted on 0 to -3 dB, we
could tack on an amplifier with a gain of 3 (i.e., $+9.5$ dB), as long as the input
impedance of the amplifier was much larger than the output impedance of the filter.
The amplifier shown in Figure E4.8 would do the job nicely.

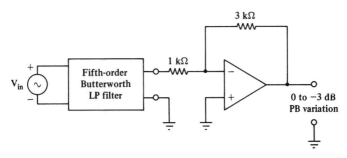

Figure E4.8

Example 4.9

What are the passband decibel levels of Example 4.6, and how could we alter them to get a filter that has a gain of −10 dB at dc?

Solution   By shorting inductors and opening capacitors, we see that the circuit behaves as two series resistors at dc.   By VDR, we get a dc gain of $1/(1 + 0.5) = 0.667$, or −3.5 dB.   Since this is an even-order Chebyshev filter, the dc level is at the bottom of the passband.   This means that the passband levels are between −2.5 and −3.5 dB.   To lower the dc level to −10 dB, we could use an amplifier with a negative gain of −6.5 dB (use 1 k-Ω and 470 Ω instead of the 1 and 3k-Ω of Example 4.8); but when lowering the gain, an even easier method is to "split" the 1-Ω load resistor into two resistors to get the 6.5-dB attenuation, as shown in the circuit in Figure E4.9. The circuit is identical to that of Example 4.6, except that we are using only 47% of the output voltage.

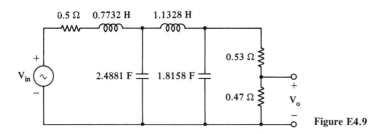

Figure E4.9

So far we have used examples that have passband variations that agree with those shown in the tables that we have available.   In the case of the Chebyshev and elliptic filters, this is necessary.   But in the case of the Butterworth, we can solve for filters even if the passband variation is different from 3 dB.   Consider Figure 4.15, which shows a specification for a Butterworth filter that has only a 2-dB variation in the passband.   Omega, Ω, is $2/0.875 = 2.28$. $A_{PB} = 2$ dB and $A_\Omega = 12$ dB.   From the Butterworth nomograph, we see that a second-order filter is required.   Butterworth filters are in Tables 4.1 to 4.3.   However, they are not for 2-dB filters that cut off at 0.875 rad/s.   But by looking at the specification, we

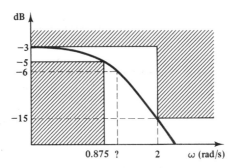

Figure 4.15

see that perhaps by the time the curve reaches 1 rad/s, it might be 3 dB down from the start level. Luckily, we have an equation that will find the frequency at which the gain is 3 dB below the dc gain of any Butterworth filter if we know the order of the filter and another frequency and the gain at that frequency relative to the dc gain. This equation is

$$\omega_{3\text{dB}} = (\omega_{\text{spec}})(10^{\text{dB}/10} - 1)^{-1/2n} \qquad \text{rad/s} \tag{4.1}$$

where ω_{spec} is the frequency specified in the problem in rad/s, dB the drop in the passband in the problem at ω_{spec}, and n the order of the filter (from nomograph). Substituting values from Figure 4.15, we get $\omega = 1$ rad/s as the point at which the curve is 3 dB down from the dc level. This means that the table values could be used directly, except for the possible addition of an op-amp or load resistor "splitting" procedure, as mentioned in Example 4.9.

Of course, it is not usually the case that the required filter will be 3 dB down for the Butterworth filter at such a low frequency as 1 rad/s. But in any case, we need to find the 3-dB frequency before we can take any Butterworth specification any further toward solution. In the next section we will find a way of adjusting the Butterworth table values (prototype values) to give us a filter that is 3 dB down at the frequency calculated from Eq. 4.1. This technique will also apply to Chebyshev, Bessel, and elliptic filters, although Eq. 4.1 is not used for these.

4.4 FREQUENCY SCALING TO SHIFT THE PROTOTYPE RESPONSE

By now, you may be saying to yourself: "Where am I going to get such large capacitors as in the previous examples, and what use would I have for a filter that cuts off at 1 rad/s (= 0.159 Hz)?" Actually, these two questions are related. It turns out that the procedure that will scale up the prototype filter into a filter that cuts off at a much higher frequency will automatically give us a reasonably sized capacitor. The only drawback is that it might then make the inductors somewhat small; but we can take care of this inductor problem in Section 4.5.

Since a small capacitor has the same effect on a circuit at a certain frequency as a larger capacitor has at a lower frequency, and since the same can be said about inductors, but not about resistors, it is reasonable to say that if we divide all capacitors and inductors by k_f, the circuit will have the same frequency response curve, except that the frequency axis will have to be relabeled with values k_f times as high.   This factor, k_f, is called the **frequency scaling factor**.

Example 4.10

Devise a Butterworth filter that will meet the specifications shown in Figure E4.10a.

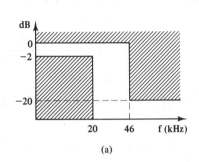

(a)

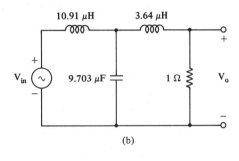

(b)

Figure E4.10

Solution   Using the Butterworth nomograph, we see that a third-order filter is required.   Because dc gain is 0 dB, Figure 4.9 should suffice.   Before we go any further, we must find the frequency at which our desired filter is 3 dB down from the dc level, by using Eq. 4.1.   This yields $\omega_{3dB} = (2\pi 20{,}000)(10^{2/10} - 1)^{-1/6} = 137{,}413$ rad/s. Thus we see that our filter is to be -3 dB at a frequency (137,413 rad/s) that is 137,413 times as high as the prototype filter (1 rad/s), so that means that $k_f = 137{,}413$.   Notice that we must not use the given 2-dB frequency for this comparison; and, of course, we must compare rad/s with rad/s, or Hz with Hz.   Now, dividing all capacitors and inductors in the prototype circuit by 137,413 yields the final circuit shown in Figure E4.10b (the prototype was taken from Example 4.1).

4.5 IMPEDANCE SCALING TO GET PRACTICAL COMPONENT VALUES

Doing the frequency scaling of the preceding section will get us a filter that will meet our specifications.   However, the resistors will be in the neighborhood of 1 Ω, which is usually impractical.   We can raise the resistor values by any factor we like and still not affect the frequency response curve if we also raise the impedance of the capacitors and inductors by the same factor; this means: multiply the inductors and divide the capacitors by the same factor that we multiplied the resistors by.   This factor, k_m, is called the impedance scaling factor.

In a circuit in which $G(s) = V_o/V_{in}$, the $G(s)$ is unchanged by this maneuver. If $G(s) = V_o/I_{in}$, then $G(s)$ is changed, but by the constant factor, k_m, uniformly over the entire frequency range; that is, the shape of the plot and the cutoff frequencies are unchanged; just the decibel scale on the graph is shifted [but not if $G(s)$ is defined as voltage out divided by voltage in].

Thus frequency scaling is important to get the response located at the proper frequency and impedance scaling is just a convenience so that standard-sized parts can be used. These two techniques are usually done in the following order: first frequency scale the circuit, then impedance scale it.

Example 4.11

Design an elliptic filter to meet the specification shown in Figure E4.11a if the source resistance is 600 Ω.

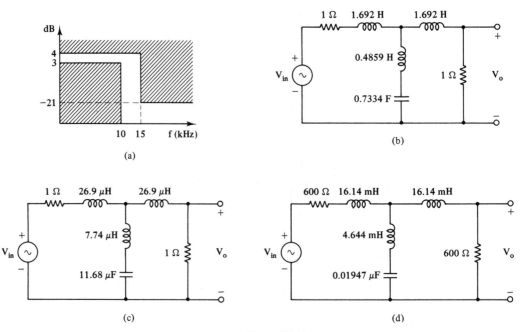

Figure E4.11

Solution

$$\Omega = 15\ \text{k}\Omega/10\ \text{k}\Omega = 1.5 \qquad A_{PB} = 4 - 3 = 1\ \text{dB} \qquad A_\Omega = 4 - (-21) = 25\ \text{dB}$$

Using these values with the elliptic nomograph yields $n = 3$. Entering the 1-dB elliptic table (Table 4.5) at $n = 3$ and $\Omega = 1.5$ gives us the prototype circuit shown in Figure E4.11b if we pick Figure 4.13. Notice that the table confirms that the stopband starts at 25.176 dB down from maximum (see A_Ω column in the table). Since we are comparing a 1-dB prototype with a 1-dB specification, we just compare the specification cutoff frequency with the table's 1 rad/s and we see that the frequency scaling factor is $2\pi10,000 = 62,800$. Dividing all L and C by 62,800 will give us the circuit shown in Figure E4.11c. To get the source resistance to 600 Ω means an impedance scaling factor of $k_m = 600$. Multiplying R and L by 600 and dividing C by 600 will then yield the final circuit shown in Figure E4.11d. This circuit will be correct in every respect, except for the overall decibel level, which could be corrected by an amplifier stage at the end, as mentioned previously.

Reference was made in Section 3.6 to the connection between the number of half-ripples and the order of the filter. If we consider an LP or a HP Chebyshev filter, it is true that the number of half-ripples equals the order of the filter. In the case of the LP and HP elliptic filter, the number of half-ripples in the passband is always equal to the number of half-ripples in the stopband, and if the order is odd, it is also equal to the order. It is one less than the order if the order is even. The reason for this is as follows. It is possible to make elliptic filters in which the order is always equal to the number of half-ripples in the passband, but in the case of even-order filters, this would require either "negative" inductors or transformers. There is a way to get around this problem, by designing the so-called case C elliptic filter for these even-order cases. These can be built with "normal" R, L, and C components, and this is what we have done for the values shown in Tables 4.4 and 4.5. The side effect is that these even-order elliptic filters have one less half-ripple. They also do not drop into the stopband quite as quickly as they would if we could use the negative inductors or transformers discussed above, but they still are a big improvement over the Chebyshev or Butterworth filters in this regard. What this all means is that if you know the order of the elliptic filter, you can immediately tell how many half-ripples it will have, but if you know the number of half-ripples you cannot tell the order of the filter, exactly. When it comes to BP filters, the number of half-ripples in the passband is exactly double what it would be for the LP case, but each stopband would have the same number as the LP case; but since there are two stopbands, this means that there are really twice as many there also. The same argument applies to the BR case, except, of course, there the stopbands are together and the passbands are separated.

Example 4.12

An elliptic LP filter designed with "normal" components has five half-ripples in the passband. What is its order, and how many half-ripples are in the stopband? Sketch the answer.

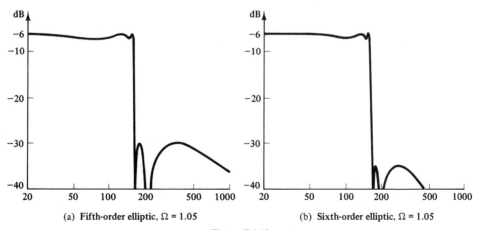

(a) **Fifth-order elliptic, $\Omega = 1.05$**   (b) **Sixth-order elliptic, $\Omega = 1.05$**

Figure E4.12

Solution It also has five half-ripples in the stopband, but the order of the filter could be either fifth (Figure E4.12a) or sixth (Figure E4.12b), since fifth- and sixth-order elliptic filters both have five half-ripples.

Example 4.13

How many half-ripples does an elliptic fourth-order BR filter have? Sketch the frequency response.

Solution If it were a LP filter, it would have one less half-ripple, since it is elliptic and is even: thus three half-ripples in the passband and three in the stopband. But since it is a BR filter, it has three half-ripples in each passband and six half-ripples in the center band, which is the stopband (see Figure E4.13).

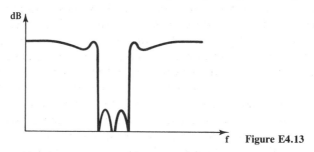

Figure E4.13

4.6 CONVERTING LOW-PASS TO HIGH-PASS, BANDPASS, AND BAND-REJECT FILTERS

Now that we have examined LP filters in detail, it remains to consider HP, BP, and BR filters. Obviously, Figure 4.16a is a LP filter and Figure 4.16b is an HP filter. The inductor and capacitor have the same impedance at $\omega = 1$ rad/s, namely 1 Ω. This makes the output at $\omega = 1$ rad/s the same for both circuits (see Figure 4.16c). At other frequencies, their impedances are different from each other, so that the two circuits have different outputs (see Figure 4.16c). However, the capacitor has the same impedance at 2.0 rad/s as the inductor has at 0.5 rad/s, making the output of the two circuits the same, but at different frequencies. Thus there is a symmetry between the two curves about the $\omega = 1$ rad/s line, as long as we use a log frequency plot. From this we can draw the general conclusion that replacing all inductors by capacitors of reciprocal value and all capacitors by inductors of reciprocal value will lead to a circuit that has a frequency plot that is the mirror image (about $\omega = 1$ rad/s) of the original circuit. This, of course, means that a LP filter can become a HP filter by this technique, and vice versa, but that a BP filter stays a BP filter and a BR filter stays a BR filter. Thus a 2-H inductor, which has an impedance of $2s$ in a LP filter, will become a 0.5-F capacitor with an impedance of $2/s$ when this circuit is transformed to a HP filter. Since $2s$ is replaced by $2/s$, this is the same as saying that if we replace all s in the

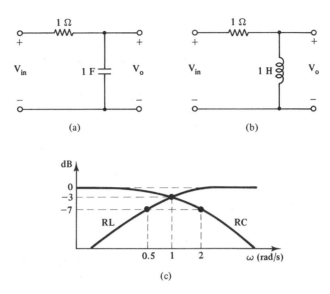

Figure 4.16

transfer function of the LP filter with $1/s$, we will have the transfer function for a HP filter.

Example 4.14

If a first-order LP filter has $G(s) = 1/(s + 1)$, what is the $G(s)$ for the corresponding HP filter?

Solution Substituting $1/s$ for s yields

$$G_{HP}(s) = G_{LP}(1/s) = \frac{1}{1/s + 1} = \frac{s}{s + 1}$$

We now have a way of making prototype HP filters from prototype LP filters. This means that we do not need a set of extensive figures and tables for the high-pass case. After the HP prototype is found, we can then frequency scale and impedance scale it just as we did for the LP case. In fact, the conversion to the HP prototype and the subsequent frequency scaling can be done all in one step. Table 4.7 shows what each inductor and capacitor in the LP prototype should be replaced by and also how to calculate its value if we want a HP filter at some cutoff frequency other than 1 rad/s. This "other" frequency is called ω_0 in Table 4.7 and is, in effect, the frequency scaling factor, k_f. The use of this table should be made clear by the following example.

Example 4.15

Find a Chebyshev filter to satisfy the specification shown in Figure E4.15a.

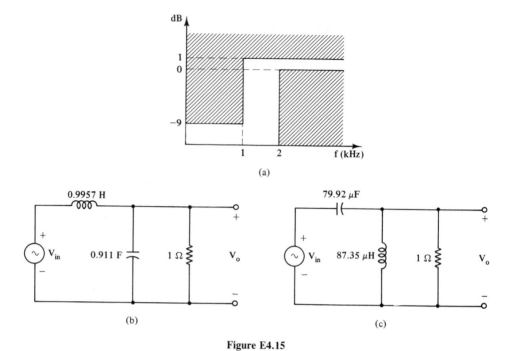

Figure E4.15

TABLE 4.7 REPLACEMENT ELEMENTS FOR CONVERTING TO HIGH-PASS, BAND-PASS, OR BAND-STOP

Given LP prototype elements	Replacement high-pass elements	Replacement bandpass elements	Replacement band-stop elements
L_P	$\dfrac{1}{\omega_o L_P}$	$\dfrac{L_P}{BW} \qquad \dfrac{BW}{\omega_o^2 L_P}$	$\dfrac{BW\,L_P}{\omega_o^2}$ $\dfrac{1}{BW\,L_P}$
C_P	$\dfrac{1}{\omega_o C_P}$	$\dfrac{BW}{\omega_o^2 C_P}$ $\dfrac{C_P}{BW}$	$\dfrac{1}{BW\,C_P} \qquad \dfrac{BW\,C_P}{\omega_o^2}$

Note: To use this table, BW and ω_o must be in rad/s.

Solution   For a HP filter, we can use the nomographs as we did before, except that now Ω = passband frequency divided by stopband frequency (otherwise, Ω would be <1, and the nomograph could not be used). So $\Omega = 2$; $A_{\text{PB}} = 1$ dB, $A_\Omega = 10$ dB, and working the Chebyshev nomograph yields $n = 2$. Picking Figure 4.9a as a suitable topology yields the LP prototype circuit shown in Figure E4.15b. Now we can convert this to a HP filter and frequency scale it all in one step by using the rules of Table 4.7, and noting that $k_f = 2\pi f_0/1$ since the prototype has a 1-dB cutoff of 1 rad/s and the desired filter has a cutoff of 2000 Hz, or 12,566 rad/s. Thus Table 4.7 will give us the HP filter circuit shown in Figure E4.15c.

Table 4.7 can also be used to convert LP to BP or BR filters; however, as before, we must find the order by using the appropriate nomograph. A_{PB} and A_Ω can be defined as before, but we will have a problem defining Ω. If the specification is as shown in Example 4.16, then is $\Omega = f_4/f_2$, or is $\Omega = f_1/f_3$? Actually, in order for the given nomographs to work for BP or BR filters, neither of these ratios will work. We will need a new definition, as follows:

$$\Omega = \frac{f_4^2 - f_0^2}{f_4(f_2 - f_1)} = \frac{\omega_4^2 - \omega_0^2}{\omega_4(\omega_2 - \omega_1)} \qquad \text{(BP and BR only)} \qquad (4.2)$$

$$f_0 = \sqrt{f_2 f_1} \quad \text{or} \quad \omega_0 = \sqrt{\omega_2 \omega_1} \qquad \text{(BP and BR only)} \qquad (4.3)$$

$$\text{BW} = \omega_2 - \omega_1 \;\text{(for BP)} \qquad \text{and} \qquad \text{BW} = \omega_4 - \omega_3 \;\text{(for BR)} \qquad (4.4)$$

It is important that f_0, the center frequency, be calculated as shown in Eq. 4.3, never as the arithmetic midpoint of f_1 and f_2. The calculation in Eq. 4.3 is called the "geometric" midpoint of the two frequencies. For any BP or BR filter that we design, it will turn out that the geometric mean (or midpoint) from Eq. 4.3 will be the same as the geometric mean of f_1 and f_2, or, as a matter of fact, of any two frequencies that have the same decibel level. This may not be true for the given specifications, but it will be true for the actual filter results.

Example 4.16

A BP filter that is designed by methods contained in this book will yield the frequency response shown in Figure E4.16. Find frequency f_3'.

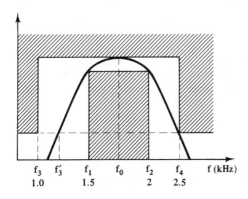

Figure E4.16

Solution   By Eq. 4.3, $f_0 = \sqrt{f_2 f_1} = 1732$ Hz.   According to the rules stated above, since f_3' and f_4 are at the same decibel level, their geometric mean must also be 1732 Hz $= \sqrt{f_3' f_4} = \sqrt{2500 f_3'}$.   Therefore, $f_3' = 1200$ Hz.   Thus we have room to spare between the 1000 Hz of the specification and the 1200 Hz of the actual filter.   Also, because of the geometric ratio rule stated previously and because of the properties of the log frequency plot, any BP or BR filter that we design will have perfect symmetry about f_0 when we view the plot.

Example 4.17

Design a Chebyshev filter to fit the specs shown in Figure E4.17a and sketch the frequency response.

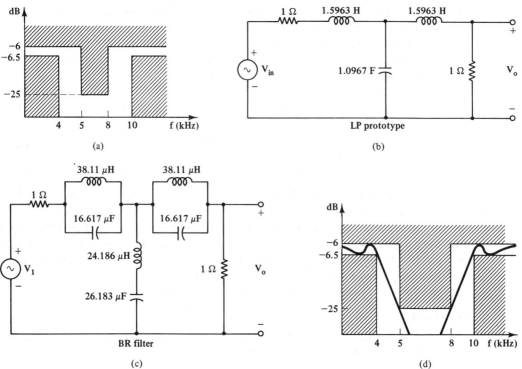

Figure E4.17

Solution   The center frequency is calculated as $f_0 = \sqrt{f_2 f_1} = \sqrt{8 \text{ k}\Omega(5 \text{ k}\Omega)} = 6325$ Hz, or as $\sqrt{f_4 f_3} = \sqrt{10 \text{ k}\Omega(4 \text{ k}\Omega)} = 6325$ Hz.   These are identical results showing that the specification is a balanced specification.   The value of omega to be used in the Chebyshev nomograph is calculated as

$$\Omega = \frac{(10 \text{ k}\Omega)^2 - (6325)^2}{10 \text{ k}\Omega(8 \text{ k}\Omega - 5 \text{ k}\Omega)} = 2.0 \qquad A_{PB} = 0.5 \text{ dB} \qquad A_\Omega = 19 \text{ dB}$$

From these values the nomograph yields $n = 3$.   Since -6-dB gain is required at dc, a topology with equal source and load resistors would be most convenient.   This would give a gain of $\frac{1}{2}$, or -6 dB at dc, since it is just a voltage division between two equal resistors because all the other components disappear at dc.   We shall pick Figure 4.11b for odd order and Table 4.2.   For this combination, the LP prototype circuit is as shown in Figure E4.17b.   Since $f_0 = 6325$ Hz and BW $= 6000$ Hz, this converts to $\omega_0 = 39{,}738$ rad/s and BW $= 37{,}700$ rad/s.   Using these values and the values from the circuit shown in Figure E4.17b and the replacement elements from Table 4.7, we get the final circuit shown in Figure E4.17c.   The sketch of the frequency response will be as shown in Figure E4.17d.   Note that for the Chebyshev response, the ripples remain in the passbands; they do not move to the stopband.

Example 4.18

Design a Chebyshev filter to fit the specs shown in Figure E4.18a, and sketch the frequency response.

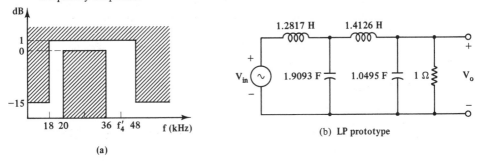

(a)

(b) LP prototype

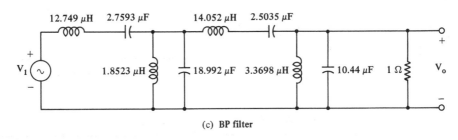

(c) BP filter

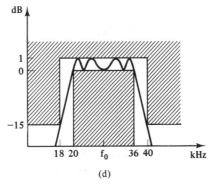

(d)

Figure E4.18

Solution   Calculating f_0 from $\sqrt{f_1 f_2}$, we get $\sqrt{20\text{ k}\Omega(36\text{ k}\Omega)}$ = 26,833 Hz.   Calculating f_0 from $\sqrt{f_3 f_4}$, we get $\sqrt{18\text{ k}\Omega(48\text{ k}\Omega)}$ = 29,394 Hz.   Thus we get different results, indicating a nonsymmetrical specification.   Since there is more room between f_2 and f_4 than we need, we can raise f_2 or lower f_4.   Let us lower f_4.   The new f_4' becomes $26,833^2/18,000$ = 40 kHz. A_{PB} = 1.0 dB and A_Ω = 16 dB.   Therefore,

$$\Omega = \frac{(40\text{ k}\Omega)^2 - 26,833^2}{40\text{ k}\Omega(36\text{ k}\Omega - 20\text{ k}\Omega)}$$

Using these values with the Chebyshev nomograph gives us n = 4.   Since it is an even-order Chebyshev, it begins at the bottom of the ripple band.   Now, since the specification calls for 0 dB at the bottom of the ripple band, it would be best to use Table 4.1 (no source resistor), since this would also produce 0 dB at dc.   We shall use Figure 4.9a.   After inserting values (for the 1-dB case) we will get the LP prototype filter shown in Figure E4.18b.   Now we can use Table 4.7 to convert to the BP case and at the same time perform the frequency scaling, by noting that $\omega_0 = 2\pi f_0$ = 168,600 rad/s and BW = $2\pi(36\text{ k}\Omega - 20\text{ k}\Omega)$ = 100,530 rad/s.   The final circuit is as shown in Figure E4.18c.   The frequency response is sketched in Figure E4.18d.

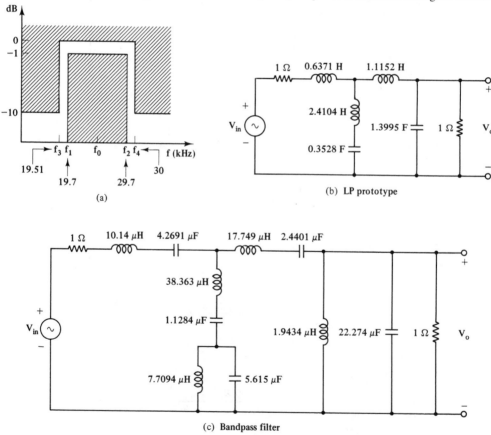

(a)

(b) LP prototype

(c) Bandpass filter

Figure E4.19

Example 4.19

Design an elliptic filter to fit the specs shown in Figure E4.19a if the source resistance is 600 Ω. Sketch the frequency response.

Solution   The ratios f_4/f_2 and f_1/f_3 both equal 1.01, so that the specification is symmetric. The center frequency $= f_0 = \sqrt{f_2 f_1} = \sqrt{29.7 \text{ k}\Omega(19.7 \text{ k}\Omega)} = 24.19$ kHz. The bandwidth $= 29.7$ kΩ $- 19.7$ kΩ $= 10$ kHz, so

$$\frac{(30 \text{ k}\Omega)^2 - (24.19 \text{ k}\Omega)^2}{30 \text{ k}\Omega(10 \text{ k}\Omega)} = 1.05 \qquad A_{\text{PB}} = 1 \text{ dB} \qquad A_{\Omega} = 10 \text{ dB}$$

Entering these values into the elliptic nomograph brings us between $n = 3$ and $n = 4$, so we will need a fourth-order filter. Using Table 4.5 and Figure 4.13 (with even-order termination) gives us the LP prototype filter shown in Figure E4.19b. Converting the above values, we have $\omega_0 = 151{,}990$ rad/s and BW $= 62{,}832$ rad/s. Using these values to convert the prototype gives us the frequency-scaled BP filter shown in Figure E4.19c. A sketch of the frequency response is shown in Figure E4.19d (note that the number of half-ripples is 3, 6, and 3).

To get a 600-Ω source impedance, we must impedance scale the circuit by letting $k_m = 600$. Multiplying all inductors and resistors by 600 and dividing all capacitors by 600 will yield the final filter, except that it will have a -6-dB gain at f_0. So as a

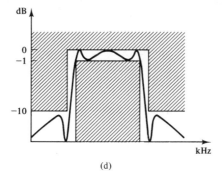

(d)

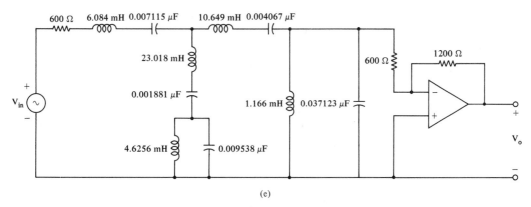

(e)

Figure E4.19 (Continued)

a last step, we can attach an amplifier with a gain of 2, as long as its input impedance is much higher than 600 Ω; or, we can use an amplifier that has exactly 600 Ω input impedance in place of the 600-Ω load resistor. Choosing the second alternative gives us the final circuit shown in Figure E4.19e.

4.7 USING THE COMPUTER TO DESIGN BUTTERWORTH AND CHEBYSHEV FILTERS

Nowadays, computers have become an invaluable aid in designing filters. The component values found in Tables 4.1 to 4.6 were all found with the computer. They could have been found by ordinary hand calculation also (except for the elliptic filters), but we have not included the formulas for finding these values. The elliptic values are quite complicated to find and would be practically impossible without a computer. For the student's use, we have included a program in Appendix B called BU/CH FILTER, which will calculate the values of all the components for any Butterworth or Chebyshev filter of any order as long as it has resistors on both ends. The computer will come up with resistors that will make the power transfer ratio a maximum. Normally, this will mean that the resistors will be of equal value, except in the case of even-order Chebyshev filters. The computer will design for any frequency level, impedance level, and any of the four responses: LP, HP, BP, or BR. It is only required that you find the order first, by consulting the proper nomograph. The program is self-explanatory. In addition to printing the values, it also prints the schematic diagram.

Example 4.20

Design a fourth-order Chebyshev BR filter, having 600-Ω impedances at each end, a BW of 2000 Hz, a center frequency of 10 kHz, and a passband ripple of 2.0 dB. Use the computer program called BU/CH FILTER. Then enter these components into the NETWORK program to see if it indeed is a fourth-order Chebyshev BR filter.

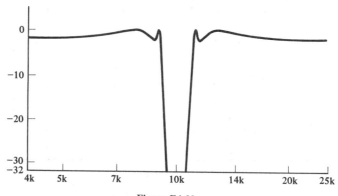

Figure E4.20

Solution Below is a listing of the input and the output values for this filter using the BU/CH FILTER program. Figure E4.20 shows the plot of the frequency response of the filter using the NETWORK program, using a start frequency of 4 kHz, an end frequency of 25 kHz, and an increment of -199, which means 199 log-spaced frequencies. Because some very large negative decibel levels might be encountered in the stopband, we have answered -33 when the program asks if we want to put a lower limit on the plot.

```
ENTER BUTTERWORTH OR CHEBYSHEV BY TYPING 'BU' OR 'CH': ? CH
TYPE 'LP', 'HP', 'BP', OR 'BR' TO INDICATE TYPE OF RESPONSE. BR
LOAD RESISTANCE (IN OHMS) = 600
ORDER = 5
ENTER CENTER FREQUENCY IN HZ: 10000
ENTER BANDWIDTH IN HZ: 2000
ENTER VARIATION IN PASSBAND, IN DB: 2
CIRCUIT VALUES ARE IN OHMS, HENRIES, AND FARADS
THIS IS A CH, BR, 2 DB, FILTER (ORDER 4)
THE TOPOLOGY AND VALUES ARE SHOWN BELOW.
(HIT ANY KEY TO CONTINUE THE OUTPUT).
SINCE THIS IS AN EVEN-ORDER CHEBYSHEV,
THE TERMINATION RESISTANCES CAME OUT
UNEQUAL.
```

SUMMARY

In this chapter we have dealt with the design of passive filters. It is assumed that the filter requirement is given; that is, we know how good the filter should be at various frequencies in terms of its gain or at least a range of allowable gain in the various frequency ranges. This statement can be put in terms of a Bode plot specification.

The choice of Butterworth, Chebyshev, or elliptic is usually quite arbitrary on the designer's part. Once decided, nomographs are used to determine the minimum order necessary. Then a circuit topology is picked out. The one used depends on whether the circuit must accommodate a source resistance or not and whether that source is a current source or a voltage source. These items are usually known before starting the problem. At this point, component values can be taken from the various tables that apply to the specific problem. This yields a so-called "prototype low-pass circuit."

Several adjustments to this prototype circuit must be made. One is to convert it to a HP, BP, or BR, if that is required. Another is to convert it so that its frequency response is shifted to a different frequency than the prototype would have given. A third adjustment would affect the impedance level, so that it would match that of other circuitry connected to this filter.

Finally, a computer program could be used to check the frequency response by inserting the found element values and observing the computer's Bode plot output and comparing it with the original specification. A computer program can also be used to do the complete design of any Butterworth or Chebyshev filter.

PROBLEMS

Sections 4.1 and 4.2

4.1. Draw a "wall" specification for the following filter: a maximum of 2-dB variation in the passband, which extends from 12 to 24 kHz; a maximum gain of 5 dB; below 5 kHz and above 40 kHz, the gain must be at least 12 dB below the maximum gain.

4.2. A filter must fit the specification shown in Figure P4.2. What is the minimum order possible, if it is a Butterworth filter?

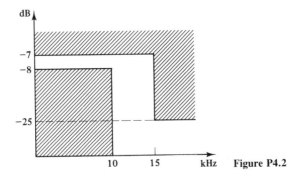

Figure P4.2

4.3. Repeat Problem 4.2 if it is to be a Chebyshev filter.

4.4. Repeat Problem 4.2 if it is to be an elliptic filter.

4.5. An LP fifth-order elliptic filter with 2-dB PB ripple leaves the PB at 10 kHz and enters the SB at 15 kHz. What is the average roll-off rate in the transition band?

Sections 4.3, 4.4 and 4.5

4.6. Realize the filter of Problem 4.2 (to within a constant) if the resistors are 1000 Ω. What is the actual dc gain? How would you adjust the circuit to get the exact decibel level shown in the specification?

4.7. Repeat Problem 4.6 for the Chebyshev case but with no source resistor.

4.8. Repeat Problem 4.6 for the elliptic case if only one inductor can be used. How much margin (in decibels) is there between the actual stopband peak and the specified stopband peak?

4.9. Realize a Butterworth third-order high-pass filter if it is to be 1.0 dB down from its peak at 10 kHz and has 1 Ω at each end.

4.10. If the load and source resistors are equal, what is their value in an LP third-order Chebyshev filter with $A_{PB} = 1.0$ dB and a cutoff of 10 kHz if the only capacitor available is 1.0 µF?

4.11. Consider the filter shown in Figure P4.11.
  (a) Is it LP, HP, BP, or BR?
  (b) Is it Butterworth, Chebyshev, Bessel, or elliptic?
  (c) Find its decibel ripple and its order.
  (d) Find its cutoff frequency.
  (e) Find its decibel level at 1 kHz.

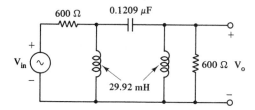

Figure P4.11

4.12. What is the order of the filter shown in Figure P4.12, and can it be realized with "normal" components?

4.13. For Figure P4.12, what is the roll-off rate in the transition band? Give your answer in dB/decade and in dB/octave.

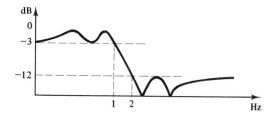

Figure P4.12

Section 4.6

4.14. The transfer function, $G(s)$, for a certain prototype LP filter is

$$G(s) = \frac{0.1253}{s^4 + 0.6115s^3 + 1.1742s^2 + 0.4318s + 0.1770}$$

Find the transfer function for the corresponding HP filter if it is to cut off also at 1 rad/s.

4.15. (a) Design a filter with ripples in the stopband and in the passband to meet the specifications shown in Figure P4.15 (within a constant) if the resistors must be 1000 Ω.
  (b) Now change the circuit so that the precise decibel levels are achieved.

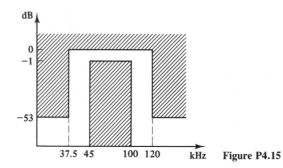

Figure P4.15

4.16. A 0.1-dB fourth-order elliptic BR filter has a stopband that is 23.736 dB down from the peak.   If the center frequency is 20 kHz and the bandwidth is 25 kHz, find f_1, f_2, f_3, and f_4.

Section 4.7

4.17. Use the BU/CH FILTER program to design a BR 3-dB Chebyshev filter using a load resistor of 600 Ω and having a BW of 1000 Hz and a center frequency of 120 kHz and order = 4.

5

analysis of active filters

OBJECTIVES

After completing this chapter, you should be able to:

- Find the transfer function, $G(s)$, from a schematic of an active circuit.
- Find the input impedance of an active circuit.
- Identify a given active filter as being MFB, Sallen-Key, or bi-quad.
- From a block diagram, set up the necessary equations to find $G(s)$, and from that $G(s)$, determine if the filter is LP, HP, BP, BR, or AP.

INTRODUCTION

Passive filters can be converted to active filters just by tacking an amplifier onto the output of a passive filter. If the amplifier has a high input impedance and a low output impedance, the advantages of this type of active filter are:

1. No loading effect; that is, the load impedance can be changed and it will not affect the output voltage or frequency response.

2. An arbitrary amount of gain is provided as a bonus; by using a variable resistor the gain can be adjusted.

3. Amplifiers are inexpensive.

These advantages are usually not enough to justify using active filters. But if we use an operational amplifier for our amplifier, then instead of just tacking the op-amp on at the end, there is a way of incorporating this op-amp into a circuit in such a way as to eliminate the inductors. This added benefit will give active filters a decided advantage, because for audio circuits, inductors are very bulky and expensive and cause magnetic fields which can be undesirable in some situations.

The disadvantages of active filters are:

1. Range of operation is usually 0 to 100 kHz and voltages and currents are limited to those specified for the particular op-amp.

2. A power supply is necessary to operate the op-amp.

These restrictions are minor and allow active filters to play an important part in radio, television, telephone, radar, satellites, and biomedical equipment.

Let us try to construct a second-order Butterworth filter with resistors, capacitors, and op-amps. We saw in Example 3.7 that a second-order low-pass filter could be constructed from two cascaded RC sections, but it is not Butterworth. Loading was a problem. If we put an op-amp between the sections, as in Figure 5.1, loading is avoided and the filter's performance improves (and even a gain of 2 is provided). However, it is still not Butterworth. In fact, it is the same response as in Example 3.8.

The point is that merely tacking on an op-amp here and there will not do the job. There are, however, several op-amp circuits that will get us any second-order transfer function we want. The op-amp circuits that we consider are the infinite-gain multiple-feedback (MFB) filter, the voltage-controlled voltage source (VCVS) or Sallen-Key filter, and the bi-quad filter, sometimes called the state-variable or universal filter.

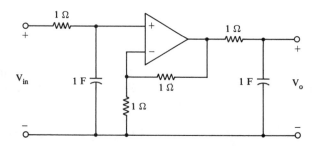

Figure 5.1

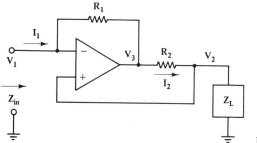

Figure 5.2

5.1 OPERATIONAL AMPLIFIER BASICS

Op-amps can also be used to create strange things, such as negative resistors, frequency-dependent negative resistors, and inductors (from resistors). For instance, it will be shown that in Figure 5.2, $Z_{in} = -Z_L$, so that if Z_L is a resistor, Z_{in} behaves as a negative resistor, meaning that input current is proportional to input voltage, but this current will flow the "wrong" way. A study of this circuit should also serve as a brush-up on the properties of op-amps in general. Ideal op-amp properties are:

1. Gain = infinite
2. Output impedance = Z_{out} = 0
3. Input impedance = Z_{in} = infinite

Because the input impedance of the op-amp is infinite, all of I_1 must flow through R_1 and all of I_2 must flow through Z_L. By VDR,

$$V_2 = V_3 \frac{Z_L}{Z_L + R_2} \qquad \text{or} \qquad V_3 = V_2 \frac{Z_L + R_2}{Z_L} \tag{5.1}$$

By Ohm's law,

$$\frac{V_1 - V_3}{R_1} = I_1 \tag{5.2}$$

Since $V_3 = (V_1 - V_2) \times$ gain and gain is infinite (or at least huge), then $V_1 - V_2$ must be zero (or very tiny), so that

$$V_1 \approx V_2 \tag{5.3}$$

Substituting Eq. 5.1 into Eq. 5.2 and replacing V_2 with V_1 we have

$$\frac{V_1 - V_1[(Z_L + R_2)/Z_L]}{R_1} = I_1 \qquad \text{or} \qquad V_1 \frac{-R_2}{Z_L R_1} = I_1$$

Therefore,

$$\frac{V_1}{I_1} = -\frac{R_1}{R_2} Z_L$$

But, by definition, $Z_{in} = V_1/I_1$, so that if $R_1 = R_2$, $Z_{in} = -Z_L$. This means that an inductor, capacitor, or resistor can be converted into a negative inductor, negative capacitor, or negative resistor. As a second benefit, by making $R_1 \neq R_2$, size scaling can be achieved. A third benefit is that a negative inductor can be achieved without using any inductor. For instance, let Z_L = a resistor and replace R_2 with a capacitor, C_2. Then

$$Z_{in} = -\frac{R_1}{1/sC_2} R_L = s(-C_2 R_1 R_L)$$

Since Z_{in} is proportional to s, it is an inductor, whose inductance value is = $-C_2 R_1 R_L$. The only restriction is that one end of this inductor must be grounded. This means that it can only be used in a circuit where the inductor is grounded anyway. This type of circuit is known as a **negative impedance converter**.

A word of caution in using negative resistors: This op-amp circuit can become unstable (driven to saturation) because of positive feedback. This will occur if this negative resistor is fed by a voltage source that has no series resistor to limit the current or that has a resistor in series with it that is smaller in magnitude than the negative resistor, thus allowing a negative current to flow. For instance, an ohmmeter contains a voltage source with a large resistance. so that it is, in effect, a constant-current source. This means that an ohmmeter can be connected to this negative resistor and it will actually read the resistance, even displaying the negative sign.

A similar circuit involving two op-amps is called a **generalized impedance converter** (GIC), which does not change the sign of the impedance but can make resistors behave as inductors or capacitors behave as resistors. From a capacitor, a new type of element can be created: a **frequency-dependent negative resistor** (FDNR), whose symbol is —|||— and whose transform impedance, Z_D, is $1/s^2D$. The FDNR is discussed in further detail in Chapter 7. Our study of active filters will use other methods to avoid the use of bulky inductors. Since a first-order filter can be designed easily enough with a passive RC circuit, we will not consider it for an active filter. However, in the next sections of this chapter we consider three active circuits for second-order filters. Higher-order filters are considered in Chapter 6 merely by cascading two or more second-order active filters, for n even. Of course, for n odd, a simple RC circuit could be added with perhaps a voltage-follower op-amp to prevent loading of the circuit.

5.2 MULTIPLE-FEEDBACK FILTER

A second-order filter with a minimum number of components, low output impedance, inverting gain, and good stability is the infinite-gain multiple-feedback filter shown in Figure 5.3. The elements are shown as admittances since that is the

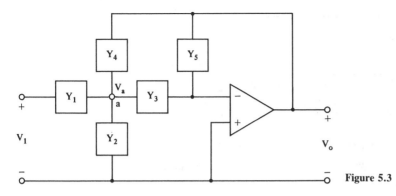

Figure 5.3

easiest form in which to handle them in the mathematical analysis of the circuit. It is called infinite gain because the voltages at the op-amp's input terminals are zero.

Writing the KCL or nodal equation at point a (where voltage is V_a), we have

$$(Y_1 + Y_2 + Y_3 + Y_4)V_a - Y_1V_1 - Y_4V_o = 0 \tag{5.4}$$

At the virtual ground at the negative terminal of the op-amp, the nodal equation is (since current into the op-amp = 0)

$$-Y_3V_a - Y_5V_o = 0 \quad \text{therefore} \quad V_a = -V_o\frac{Y_5}{Y_3} \tag{5.5}$$

Substituting Eq. 5.5 into Eq. 5.4 yields

$$(Y_1 + Y_2 + Y_3 + Y_4)\frac{Y_5}{Y_3}(-V_o) - Y_1V_1 - Y_4V_o = 0$$

or

$$V_o\left[\frac{Y_5}{Y_3}(Y_1 + Y_2 + Y_3 + Y_4) + Y_4\right] = -V_1Y_1$$

Therefore,

$$\frac{V_o}{V_1} = -\frac{Y_1Y_3}{Y_5(Y_1 + Y_2 + Y_3 + Y_4) + Y_3Y_4} \tag{5.6}$$

Since all admittances and all signs are positive, except the sign in front of the right side of Eq. 5.6, the gain is inverted. Since we agree that only resistors and capacitors will be used, the Y values must be either constants or s times a constant.

For a second-order low-pass filter, we require the numerator to be a constant. Hence Y_1 and Y_3 are resistors. Thus an s^2 term in the denominator can only be had if Y_5 is a capacitor. This means that the constant term in the denominator can exist only if Y_4 is also a resistor. This means that Y_2 must also be a capacitor, so that the s^2 term can be generated. Thus the circuit in Figure 5.4 is a second-

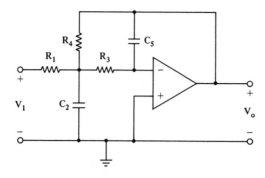

Figure 5.4

order low-pass filter and $Y_1 = 1/R_1$, $Y_2 = sC_2$, $Y_3 = 1/R_3$, $Y_4 = 1/R_4$, and $Y_5 = sC_5$. Substituting these relations into Eq. 5.6 yields

$$G_{LP}(s) = \frac{V_o}{V_1} = \frac{-1/(R_1R_3)}{sC_5[1/R_1 + 1/R_3 + 1/R_4 + sC_2] + 1/(R_3R_4)}$$

$$= \frac{-\dfrac{R_4}{R_1}\left[\dfrac{1}{R_3R_4C_2C_5}\right]}{s^2 + s\left(\dfrac{1}{C_2}\right)\left(\dfrac{1}{R_1} + \dfrac{1}{R_3} + \dfrac{1}{R_4}\right) + \dfrac{1}{R_3R_4C_2C_5}} \qquad (5.7)$$

From Eq. 5.7 we see that the gain at $\omega = 0$ (let $s = 0$) is $-R_4/R_1$.

Because the numerator contains just one term, LP, HP, and BP filters of the Butterworth and Chebyshev variety can be realized. However, inverse Chebyshev and elliptic filters, as well as all BR and AP filters, cannot be realized, since these involve more than one term in the numerator.

Example 5.1

For the MFB filter shown in Figure E5.1, is it LP, HP, and so on? What is its roll-off rate and order? Find the $G(s)$. Is it Butterworth or Chebyshev? What is the cutoff frequency?

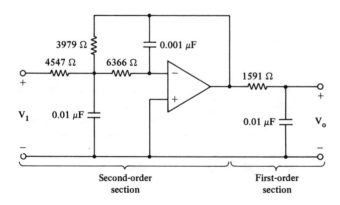

Second-order
section

First-order
section

Figure E5.1

Solution Comparing the circuit to Figure 5.4, we see that it is a second-order low-pass filter with an RC section tacked on: Hence it is a third-order LP filter, so its roll-off rate is 60 dB in the stopband.   The $G(s)$ of the quadratic section is (using Eq. 5.7)

$$G(s) = \frac{-(3979/4547)\{1/[(6366)(3979)(10^{-8})(10^{-9})]\}}{s^2 + s(1/10^{-8})(1/4547 + 1/6366 + 1/3979) + 1/[6366(3979)10^{-8}(10^{-9})]}$$

The $G(s)$ of the RC section is (by VDR)

$$G(s) = \frac{1/(10^{-8}\,s)}{1591 + 1/(10^{-8}\,s)}$$

Multiplying the numbers out and multiplying the two transfer functions yields the transfer function of the entire circuit:

$$G(s) = \frac{-3.455 \times 10^9}{s^2 + 62{,}833s + 3.9478 \times 10^9} \times \frac{1}{1 + 1.591 \times 10^{-5}\,s}$$

To find out if this is Butterworth, we must look at $|G(j\omega)|$:

$$|G(j\omega)| = \frac{0.875}{\sqrt{1 - 2.533 \times 10^{-10}\,\omega^2 + 6.416 \times 10^{-20}\,\omega^4}}$$

$$\times \frac{1}{\sqrt{1 + 2.533 \times 10^{-10}\,\omega^2}}$$

Although the first fraction is not Butterworth, we do find that when both fractions are multiplied together, the result is Butterworth, since it fits the form $G(j\omega) = k/\sqrt{1 + (\omega/\omega_0)^{2n}}$. Actually, $|G(j\omega)| = 0.875/\sqrt{1 + (\omega/62{,}832)^6}$, so that $\omega_0 = 62{,}832$ rad/s.   This means that the cutoff frequency $f_0 = 62{,}832/2\pi = 10{,}000$ Hz.   This illustrates a very important point: A higher-order Butterworth filter is never two or more Butterworth functions multiplied together. No sections are actually Butterworth, yet the total $G(s)$ is Butterworth.

Example 5.2

For the circuit in Figure E5.2a, determine the type of filter, its $G(s)$, its roll-off rate, and its cutoff frequency.

Solution This has the topology of a MFB filter; however, component 1 is now a capacitor, and component 2 is now a resistor. This is probably not a LP filter. We must substitute values into Eq. 5.6.

$$G(s) = \frac{-s(10^{-8})/10{,}058}{s(10^{-9})[s(10^{-8}) + 1/8670 + 1/10{,}058 + 1/10{,}058] + (1/10{,}058)(1/10{,}058)}$$

$$= \frac{-99{,}420s}{s^2 + 31{,}420s + 9.87 \times 10^8}$$

At low frequencies, $G(s)$ is approximately $-1.0058 \times 10^{-4}s$. This is an asymptote that is rising at a rate of 20 dB/decade. At high frequencies, $G(s)$ is approximately $-9.942 \times 10^4/s$. This is an asymptote that is falling at a rate of -20 dB/decade.

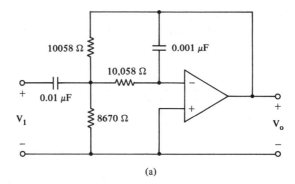

(a)

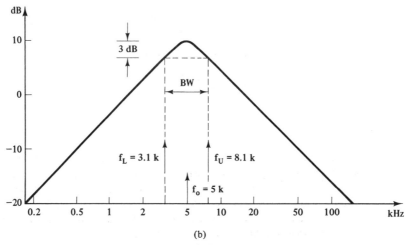

(b)

Figure E5.2

Thus the conclusion is that this is a bandpass filter that has its center frequency at the point of intersection of the two asymptotes. By equating the two expressions for $G(s)$ and solving for s, we find that the center frequency is 31,420 rad/s or 5 kHz. Substituting this value of $j31{,}420$ for s, we get a gain at resonance of 10 dB. By setting the polar form of $G(j\omega)$ = magnitude corresponding to 7 dB (3 dB down from the peak), the two cutoff frequencies can be found. Another alternative is to use the NETWORK program with the values from the schematic. Using a nominal gain of 100,000 and an op-amp output impedance of 10 Ω, the plot shown in Figure E5.2b is obtained from this program.

5.3 SALLEN-KEY FILTER

The second type of active filter we study is the Sallen-Key filter, or voltage-controlled voltage-source (VCVS), filter. Its general configuration is shown in Figure 5.5. It has finite gain in the sense that both voltages at the op-amp inputs are greater than zero.   It has only one component more than the previous filter, noninverting gain, and low output impedance.   The low-pass version can be adjusted easily; in particular, the gain can be changed by adjusting R_3 or R_6 (the two can be in the form of a potentiometer).   The op-amp gain is

$$1 + \frac{R_6}{R_3} = \mu \tag{5.8}$$

Writing nodal equations at nodes a and b gives us

$$(Y_4 + Y_1 + Y_5)V_a - Y_4V_1 - Y_5V_o - \frac{Y_1V_o}{\mu} = 0 \tag{5.9}$$

$$(Y_2 + Y_1)\frac{V_o}{\mu} - Y_1V_a = 0 \quad \text{or} \quad V_a = \frac{(Y_2 + Y_1)V_0}{\mu Y_1} \tag{5.10}$$

where we have used V_o/μ instead of V_b.
Substituting Eq. 5.10 into Eq. 5.9 and solving for V_o/V_1 yields

$$G(s) = \frac{V_o}{V_1} = \frac{\mu Y_4 Y_1}{(Y_2 + Y_1)(Y_4 + Y_1 + Y_5) - Y_1^2 - \mu Y_5 Y_1} \tag{5.11}$$

For a low-pass filter, the numerator must be a constant.   This again rules out inverse Chebyshev, elliptic, and all band-reject and all-pass filters just as it did for the MFB circuit.   Since only resistors and capacitors will be used, this means Y_1 and Y_4 are resistors.   Thus the necessary s^2 term in the denominator can only occur if Y_2 and Y_5 are capacitors.   Thus, for low-pass second-order filters,

$$G_{\text{LP}}(s) = \frac{\mu/(R_4 R_1)}{(sC_2 + 1/R_1)(1/R_4 + 1/R_1 + sC_5) - 1/R_1^2 - \mu sC_5/R_1}$$

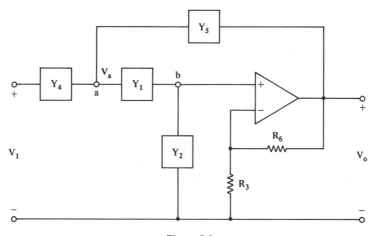

Figure 5.5

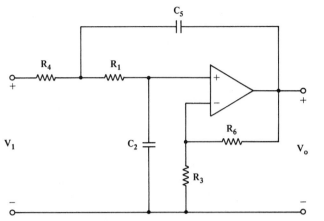

Figure 5.6

Dividing numerator and denominator by C_2C_5 and rearranging, we have

$$G_{LP}(s) = \frac{\mu/(R_4R_1C_2C_5)}{s^2 + s\left[\dfrac{1/R_4 + 1/R_1}{C_5} + \dfrac{(1/R_1)(1 - \mu)}{C_2}\right] + 1/(R_4R_1C_2C_5)} \qquad (5.12)$$

The low-pass circuit is shown in Figure 5.6. Substituting $s = 0$ into Eq. 5.12, we find the overall dc gain of the circuit to be μ, which is the same as the op-amp gain given by Eq. 5.8.

In passive filters of higher order, tuning is extremely difficult due to the interaction of all the elements. However, as we shall see later, higher-order active filters will be a cascading of two or more second-order active filters such as described above. Because of the low output impedance of each op-amp, the sections do not affect each other. Thus each second-order section can be tuned independently.

5.4 BI-QUAD FILTER

Until now, the transfer function was derived by writing nodal equations for various circuits. These were then solved and a transfer function was obtained. The coefficients of the s terms bore no direct relation to the circuit components. In the bi-quad filter, which is sometimes called the state-variable filter, a more direct relation exists between components and transfer function. In fact, block diagrams such as were used in Section 2.4 are most appropriate. Consider the circuit of Figure 5.7a. The nodal equation at point a is (since $V_a = 0$)

$$\frac{V_1}{R_1} + V_o sC = 0,$$

or

$$\frac{V_o}{V_1} = G(s) = \frac{-1}{sC_2R_1}$$

If R_1 is chosen to be equal to $1/C_2$, then

$$\frac{V_o}{V_1} = -\frac{1}{s} \qquad (5.13)$$

Thus the circuit can be represented by the block diagram of Figure 5.7b, and it is called an **inverting integrator**. Differentiators could also be produced, but they tend to be noisier. Since integrators tend to smooth out bumps and spikes of the input, we will be using just integrators and summing amplifiers.

Consider now the transfer function of a second-order low-pass Butterworth filter:

$$G(s) = \frac{1}{s^2 + \sqrt{2}\,s + 1} = \frac{V_o(s)}{V_1(s)} \qquad (5.14)$$

Removing fraction bars (cross-multiplying) gives

$$V_1(s) = V_o(s)(s^2 + \sqrt{2}\,s + 1)$$

or (5.15)

$$s^2V_o = V_1 - \sqrt{2}\,sV_o - V_o$$

If voltage V_o were available, sV_o could be generated by differentiating V_o, and s^2V_o could be generated by differentiating sV_o. Or, if s^2V_o were available, sV_o could be generated by integrating s^2V_o, and V_o could be generated by integrating sV_o (see Figure 5.8). As mentioned before, integration is the preferred method. Actually, neither s^2V_o nor V_o is available, but by the magic of feedback, s^2V_o can

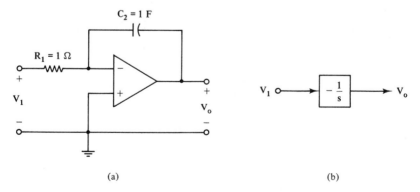

(a) (b)

Figure 5.7

Figure 5.8

be generated by using Eq. 5.15 and an adder, an amplifier, and an inverter. Studdents sometimes ask: "How is it possible to generate s^2V_o by adding various signals, if one or more of these signals ($-\sqrt{2}\, sV_o$ and $-V_o$) depend on s^2V_o for their existence?" But this is no different than asking how can $x = 2 - x$ be solved for x unless we already know x (for insertion into the right side of the equation). The feedback circuit of Figure 5.9 provides the bootstrapping operation that makes this possible, and it does not take several "cycles" before it "works." The student is encouraged to look at the circuit and do the addition or multiplication at each box to see that this circuit does satisfy Eq. 5.15. Since all the boxes and circles represent op-amps, there is no loading effect, and the signal at any point can be used as an input to any other circuitry to which we may want to apply it. The easiest summing amplifier is the inverting type. In this case the summing amplifier and integrator can be combined into one unit (see Figure 5.10, or dashed line box of Figure 5.9). Instead of using an amplifier to operate on $-sV_o$, the 0.707-Ω resistor in Figure 5.10 has the same effect, since the current into the node at the op-amp's virtual ground is

$$I = \frac{-sV_o}{0.707} + \frac{-V_o}{1} + \frac{V_1}{1}$$

$$= V_1 - V_o - \sqrt{2}\, sV_o$$

If the feedback element were a 1-Ω resistor, the op-amp's output voltage would be $-I$ times 1 Ω, or $V_o - V_1 + \sqrt{2}\, sV_o$, which is just the negative of what we should have coming out of the adder of Figure 5.9. But since we are using a

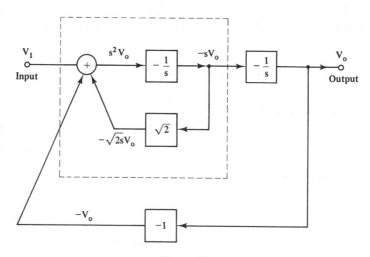

Figure 5.9

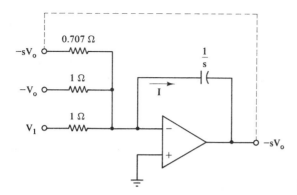

Figure 5.10

1-F capacitor, the output is $1/s$ times the output above, or exactly what it should be coming out of the first inverting integrator. This means that this output should be fed back to the 0.707 resistor, as shown by the dashed line in Figure 5.10. All that is necessary is to tack on another inverting integrator and also an inverter in the last feedback line, which results in the final circuit shown in Figure 5.11.

Comparing the Butterworth filter of Figure 5.11 with that of Sections 5.2 and 5.3 (see Examples 5.1 and 5.2), the circuits and component values are different, yet all circuits produce the same transfer function as the passive filter of Figure 3.8b. In terms of number of components, the bi-quad filter is the most complex.

The bi-quad circuit has good stability and is the easiest to tune of the three active filters. The gain can be adjusted at any time by adjusting R_3. Another

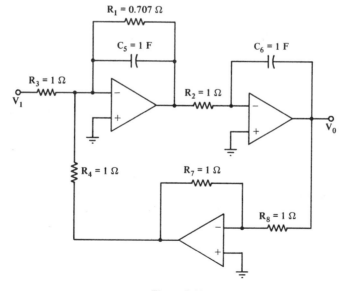

Figure 5.11

advantage is that high-Q values can be achieved with a bi-quad filter, which can be important in the design of bandpass filters.

Example 5.3

Find the component values for the circuit of Figure 5.4 if the filter function is

$$G(s) = \frac{1}{s^2 + \sqrt{2}\,s + 1}$$

Solution If we made all resistors $= 1\ \Omega$ and all capacitors $= 1$ F, we would have, from Eq. 5.7,

$$G(s) = \frac{-1}{s^2 + 3s + 1}$$

The coefficient of the s term is too high. By studying Eq. 5.7, we see that by changing C_2 to $3/\sqrt{2} = 2.12$ F, we can get the proper coefficient of s. There are, of course, other ways of accomplishing this goal. Notice that to keep this change from affecting other coefficients, C_2C_5 must stay $= 1$, meaning that $C_5 = \sqrt{2}/3 = 0.47$ F. Thus the circuit is as shown in Figure E5.3. If these values are impractical, they could be impedance scaled, making the capacitors 212 μF and 47 μF, and the resistors all 10 kΩ, for instance. After being built, this circuit would be tuned as outlined in this section. The only slight problem is that the output must be fed into an inverter to get rid of the negative sign. However, in most applications, we are not concerned about the sign.

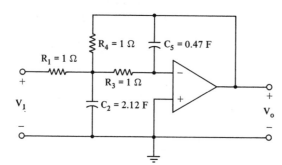

Figure E5.3

Example 5.4

Repeat Example 5.3, using the Sallen-Key circuit.

Solution The circuit to be used is Figure 5.6 and the equations are Eqs. 5.12 and 5.8. According to the discussion, we should make $R_1 \gg R_4$; but keep $R_1R_4 = 1$, so that the end term remains $= 1$. So let $R_1 = 10$ and $R_4 = 0.1$ and all the other components $= 1$. Then

$$G(s) = \frac{2}{s^2 + 10s + 1}$$

Although we can vary the gain by changing R_6 or R_3, we will never be able to get a gain of 1. This is usually no problem in a practical application. If it is, we could follow this circuit with an inverting op-amp whose gain is $-\frac{1}{2}$ followed by another whose gain is -1. The real problem again is how to adjust the coefficient of s to be $\sqrt{2}$. If we let $C_2C_5 = 1$, so as not to affect the other terms, the coefficient of s is

$$\frac{10.1}{C_5} - \frac{0.1}{C_2} \quad \text{or} \quad \frac{10.1}{C_5} - 0.1C_5 = \sqrt{2}$$

since $C_2 = 1/C_5$. Multiplying through by C_5, we have

$$-0.1C_5^2 - \sqrt{2}\,C_5 + 10.1 = 0$$

Solving this quadratic, we find that $C_5 = 5.217\,\text{F}$; therefore, $C_2 = 0.1917\,\text{F}$. Applying these new values to Eq. 5.12, we get

$$G(s) = \frac{2}{s^2 + 1.414s + 1}$$

The circuit is as shown in Figure E5.4. Again, scaling the capacitors by 10^{-4} and the resistors by 10^4 would make for more practical values.

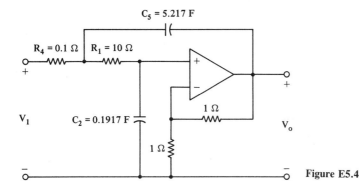

Figure E5.4

Example 5.5

Draw a block diagram for a second-order elliptic filter whose transfer function is

$$G(s) = \frac{0.0953(s^2 + 7.464)}{s^2 + 0.762s + 0.889} = \frac{V_o}{V_1}$$

Then draw its schematic, using a bi-quad filter.

Solution   Cross-multiplying to remove fraction bars gives us

$$0.0953(s^2 + 7.464)V_1 = (s^2 + 0.762s + 0.889)V_o$$

This is somewhat different than in the text material, because V_1 is also multiplied by s^2.

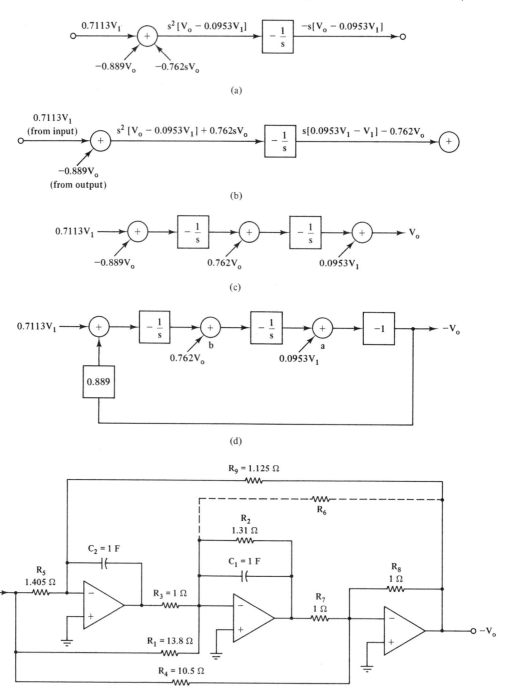

Figure E5.5

To use integrators, the left side must start with the highest power of s for V_1 as well as V_o. Thus a trial arrangement of the equation would be (Figure E5.5a)

$$s^2(V_o - 0.0953V_1) = -0.762sV_o - 0.889V_o + 0.7113\ V_1$$

This partial diagram seems okay, except where can we get $-0.762sV_o$ from? Note that the output of the integrator is $-sV_o + 0.0953sV_1$. We certainly do not want to involve sV_1 at the adder since it is not in the previous equation; and we cannot subtract $0.0953sV_1$ after the first integrator since we do not have it available. The only thing to do is rearrange the equation (Figure E5.5b):

$$s^2(V_o - 0.0953V_1) + 0.762sV_o = -0.889V_o + 0.7113V_1$$

The output of this first integrator will be fed into the second integrator, but we do not want to integrate $-0.762V_o$, so we get rid of it by adding $0.762V_o$ (from output V_o) at the second adder. After the second integrator, we must add $0.0953V_1$ (from input), leaving us with the output, $-V_o$ (Figure E5.5c).

Because adders are associated with inverting op-amps, the last adder really can be thought of as an adder and inverter (Figure E5.5d). Notice that the second adder requires a positive V_o, which is not available (unless another inverter is added). However, at point a, we do have $+V_o$ available to be used; but it unfortunately contains $-0.0953V_1$ as well. When the signal from a is multiplied by 0.762 and fed back to point b, an extra unwanted $-0.0953V_1(0.762)$, or $-0.0752V_1$, is added. This can be canceled, of course, by an extra $+0.0725V_1$ from the input and no harm is done. The final circuit is as shown below. The fact that only $-V_o$ is available is usually of no consequence.

The schematic would be as shown in Figure E5.5e. If a ninth resistor, R_6 (shown with dashed line), is added, the circuit is called an all-purpose bi-quad, because it is capable of simulating the bi-quadratic equation,

$$G(s) = -\frac{ds^2 + es + a}{s^2 + bs + c} \tag{5.16}$$

as long as b and c are positive and d and a have the same sign. It is called a bi-quadratic equation because it is a quadratic divided by a quadratic.

There are other realizations of bi-quads; some use noninverting op-amps or even just one op-amp. This concludes our study of analysis of the basic type of op-amp circuits, except to say that op-amps have limits that must be considered in any practical design, the most important of which are the gain–bandwidth product, compensation, and dc offset. In the next chapter we deal with synthesis of higher-order active filters and the adaptation of low-pass to other types: such as high-pass, bandpass, and so on.

Example 5.6

What must the values in Figure 5.11 be changed to in order to have a Bessel low-pass filter?

$$G_{\text{Be}}(s) = \frac{1}{s^2 + 3s + 3} \text{ (from Section 3.9)} = \frac{V_o}{V_1}$$

Then

$$(s^2 + 3s + 3)V_o = V_1$$

or

$$s^2 V_o = V_1 - 3sV_o - 3V_o$$

This means that the feedback ratios of $\sqrt{2}$ and -1 of Figure 5.9 are replaced by 3 and -3; or replace the R_1 and R_4 of Figure 5.11 with 0.333 Ω each.

Example 5.7

What must Figure 5.11 be changed to, in order that it is Butterworth but with a cutoff of 20 kHz and capacitors = 0.001 μF?

Solution First, make a frequency transformation: $k_f = 2\pi 20{,}000 = 125{,}600$. Therefore, capacitors = $1/125{,}600 = 7.97$ μF; the resistors stay the same. To get the capacitors = 0.001 μF, use impedance scaling to a higher impedance: $k_m = 7.97$ μF/0.001 μF = 7970. This makes $R_1 = 5630$ Ω and the other resistors = 7970 Ω, and of course, the capacitors are now 0.001 μF.

SUMMARY

In this chapter we saw how second-order low-pass filters can come about by using op-amps and resistors and capacitors. Filters such as Butterworth and Chebyshev can exist without using inductors. This is one of the big advantages of active circuits. Although many different configurations exist, only the multiple-feedback, Sallen-Key, and bi-quad forms are studied here.

PROBLEMS

Section 5.1

5.1. Find $G(s) = V_o(s)/V_1(s)$ for the active circuit shown in Figure P5.1.

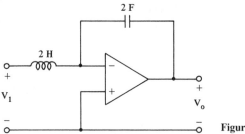

Figure P5.1

5.2. Repeat Problem 5.1 for the circuit in Figure P5.2.

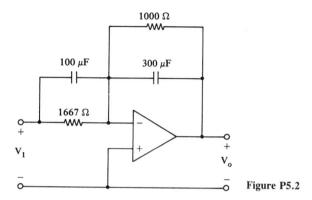

Figure P5.2

5.3. In Figure P5.3, element 1 is a short, elements 2 and 4 are 100 μF capacitors, and element 3 is a 0.01-μF capacitor.
    (a) Find a simplified expression for $Z_T(s)$.
    (b) What is the type of element represented by $Z_T(s)$ and its value?

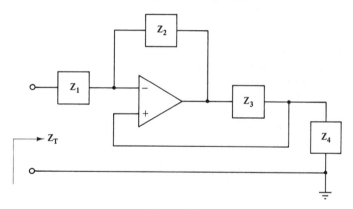

Figure P5.3

5.4. Repeat Problem 5.3 if element 1 is a short, elements 2 and 4 are 1-μF capacitors, and element 3 is a 3-Ω resistor.

Section 5.2

5.5. From Eq. 5.6, is it possible to create $G(s) = s^2/(s^2 + 3s + 4)$ just by using resistors and capacitors? If so, which are the capacitors and which are the resistors, and what are the values of the resistors if the capacitors are 1 F?

5.6. Repeat Problem 5.5 for $G(s) = s/(s^2 + 3s + 4)$ if elements 1 and 2 are known to be 1 F capacitors.

Section 5.3

5.7 Consider the circuit of Figure P5.7.
   (a) Find $G(s) = V_o(s)/V_1(s)$.
   (b) Factor $G(s)$.
   (c) Make a Bode plot of $G(s)$.

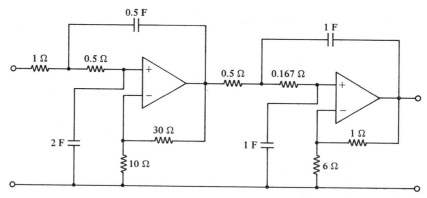

Figure P5.7

Section 5.4

5.8. Consider the circuit in Figure P5.8.   By mathematical analysis
   (a) Find the transfer function, $G(s) = V_o(s)/V_1(s)$.
   (b) Is it Butterworth or Chebyshev?
   (c) Is it LP, HP, and so on, and what is its order?
   (d) How much ripple does it have?
   (e) What is its cutoff frequency?

5.9. Repeat Problem 5.8b to e by entering circuit values into the NETWORK program and analyzing the plot.

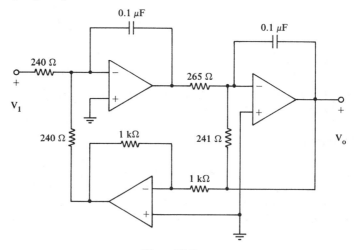

Figure P5.9

6

synthesis of active filters

OBJECTIVES

After completing this chapter, you should be able to:

- Break a given transfer function into factors containing quadratic denominator factors by using the computer.
- Find a transfer function by consulting tables of the various filter types (Butterworth, Chebyshev, etc.).
- Realize an active filter of the MFB, Sallen-Key, or bi-quad type if the quadratic factored form of $G(s)$ is given.
- Convert the given $G(s)$ for a prototype filter into an HP (or BP or BR) form.
- Realize an active HP, BP, or BR filter as one of the three types: multiple-feedback, Sallen-Key, or bi-quad.
- Use a computer program to verify a given active circuit and interpret the results of the computer plot.

INTRODUCTION

Recall that when synthesizing passive filters, we went from the Bode specification to the nomographs to find the order of the filter and then directly to tables of component values. The transfer function, $G(s)$, was never used or even calculated. When synthesizing active filters, we will also go from the Bode specification to the nomographs to get the filter order; but since it is relatively easy to synthesize a $G(s)$ for the active case, we will use tables to find the $G(s)$ and calculate the component values instead of reading them from a table. This is the basic difference in approach between active synthesis and passive synthesis. Frequency scaling and impedance scaling are similar to the passive approach, but conversion from LP to HP, and so on, is quite different in most cases. Finally, it is possible to synthesize an active all-pass filter without using the bridge circuit that was necessary in the passive case.

6.1 FACTORING THE FILTER'S TRANSFER FUNCTION

In the passive case it was difficult to synthesize higher-order filters because each section of the ladder interacted with every other section, so that we had to resort to tables of component values. We could do the same with active filters, except that the circuits are more complex and the tables would become large. Also, since any filter could be made using any of the three types (MFB, Sallen-Key, or bi-quad), the tables become even larger. Even the method of converting low-pass to high-pass, and so on, is different. There is, however, one facet of active filters that makes synthesis somewhat easier. Op-amps have the characteristic in which the sections do not load one another. That means that a fourth-order filter can be broken down into two second-order sections cascaded together; that is, the output of one will be fed as an input to the next, as shown in Figure 6.1a. An alternative method, shown in Figure 6.1b, is to have the input feed two sections, whose output is then added to give V_o.

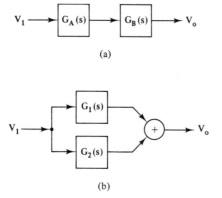

Figure 6.1

Let us consider the transfer function

$$G(s) = \frac{1}{(s^2 + 0.7654s + 1)(s^2 + 1.848s + 1)} \tag{6.1}$$

This is a fourth-order low-pass Butterworth filter, which when broken down by partial fractions is

$$G(s) = \frac{-0.9239s - 0.7071}{s^2 + 0.7654s + 1} + \frac{0.9239s + 1.7071}{s^2 + 1.8485s + 1} = G_1(s) + G_2(s) \tag{6.2}$$

Multiplying by V_1 yields

$$V_o = V_1 G_1(s) + V_1 G_2(s) \tag{6.3}$$

so that the circuit of Figure 6.1b would work. For instance,

$$G_1(s) = \frac{-0.9239s - 0.7071}{s^2 + 0.7654s + 1}$$

could be realized with the all-purpose bi-quad discussed in Chapter 5 if the proper resistor and capacitors are picked. The same goes for the realization of $G_2(s)$. However, this method of partial fraction expansion is very seldom used. Instead, we will realize the more common cascade circuit of Figure 6.1a, just by bracketing the factors as follows:

$$G(s) = \left[\frac{1}{s^2 + 0.7654s + 1} \right] \left[\frac{1}{s^2 + 1.848s + 1} \right] = [G_A(s)][G_B(s)] \tag{6.4}$$

It is just as easy to cascade sections as it is to add outputs of side-by-side sections. Besides, the bracketing operation is much simpler, and each section is a low-pass filter. Although neither section is a second-order Butterworth filter, when they are combined by cascading they produce the required fourth-order low-pass Butterworth filter.

If $G_A(s)$ feeds $G_B(s)$, and $G_A(s)$ is defined as V_{o_A}/V_{1_A} and is equated to the first brackets of Eq. 6.4, and $G_B(s)$ is defined as V_{o_B}/V_{1_B} and is equated to the second brackets of Eq. 6.4, then by cross-multiplying these new equations, we have

$$s^2 V_{o_A} = V_{1_A} - 0.7654s V_{o_A} - V_{o_A} \tag{6.5}$$

$$s^2 V_{o_B} = V_{1_B} - 1.848s V_{o_B} - V_{o_B} \tag{6.6}$$

where

$$V_{o_A} = V_{1_B} \tag{6.7}$$

Equations 6.5 and 6.6 can be realized similar to the way Eq. 5.15 was realized in Figure 5.11, yielding the circuit shown in Figure 6.2. Thus, breaking the given transfer function into groups, each having a quadratic as a denominator and then using one of the standard active forms from Chapter 5, it is fairly easy to realize

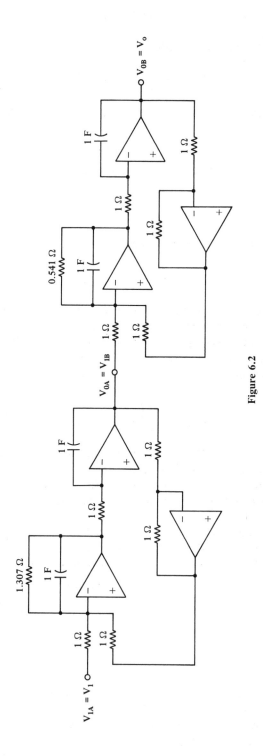

Figure 6.2

an active form of any higher-order filter. If the transfer function was already factored as in the previous discussion, most of the work is already done. If it is not factored, we must do the factoring ourselves. This is almost impossible without a computer program. So we use the program called ROOT FINDER in Appendix B. This will give us the binomial factors. To get quadratic factors, we must combine these, two at a time.

Example 6.1

Find the quadratic factors of

$$G(s) = \frac{1}{s^4 + 2.6131s^3 + 3.4142s^2 + 2.6131s + 1}$$

Solution Entering the coefficients of the denominator into the computer program called ROOT FINDER, we get

$$p_1 = -0.3827 + j0.9239 \qquad p_2 = -0.3827 - j0.9239$$

$$p_3 = -0.9239 + j0.3827 \qquad p_4 = -0.9239 - j0.3827$$

Thus the binomial factors are

$$(s + 0.3827 - j0.9239)(s + 0.3827 + j0.9239)$$

$$\times (s + 0.9239 - j0.3827)(s + 0.9239 + j0.3827)$$

There are three different ways of combining these factors two at a time. In the case of complex roots, we must put a factor with its complex conjugate so as to avoid imaginary numbers in the quadratics. In this example, that means to combine the first with the second and combine the third with the fourth, yielding

$$G(s) = \left[\frac{1}{s^2 + 0.7654s + 1} \right] \left[\frac{1}{s^2 + 1.848s + 1} \right]$$

which just happens to be Eq. 6.1.

Many times $G(s)$ will be given to us already factored into quadratics. Of course, in the case of odd-order filters there will be a binomial left over. This can be realized passively with an RC circuit, but if loaded by any following circuitry, the frequency response will change. In that case it is better to use an op-amp for this binomial section, as the next example shows.

Example 6.2

Realize the circuit whose transfer function is

$$G(s) = \frac{1}{(s + 1)(s + 2)(s + 3)} \qquad \text{(a third-order LP filter)}$$

Solution Since there are no complex poles, we can combine any two factors.

$$G(s) = \left[\frac{1}{s^2 + 5s + 6} \right] \left[\frac{1}{s + 1} \right]$$

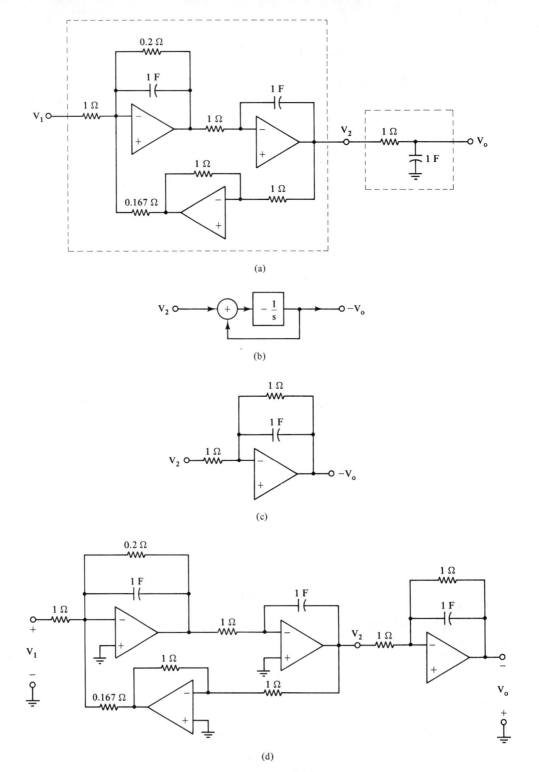

(a)

(b)

(c)

(d)

Figure E6.2

The first term can be realized as a bi-quad and the second as an *RC* section (see Figure E6.2a). To avoid loading of the preceding section, we convert it to an active circuit also. For this section

$$\frac{V_o}{V_2} = \frac{1}{s + 1} \quad \text{or} \quad V_o(s + 1) = V_2 \quad \text{or} \quad V_o s = V_2 - V_o$$

for which the block form is shown in Figure E6.2b. The op-amp form is shown in Figure E6.2c. Therefore, the 100% active form of the entire filter is shown in Figure E6.2d. The slight difference from what was asked for is that the output is inverted.

6.2 LISTING OF VARIOUS FILTER FACTORS

Instead of having to find the factors of standard filter types, we will have these factors available to us in tabular form. Thus Table 6.1 will be for a Butterworth filter, Table 6.2 for a Chebyshev, Table 6.3 for an elliptic, and Table 6.4 for a Bessel. Thus, if a seventh-order Butterworth low-pass filter is desired, we look up the factors in the table, build a bi-quad (or MFB or Sallen-Key) for each quadratic factor and a section for the odd term, and then cascade them in any order. In fact, we could make one section a bi-quad, another section a Sallen-Key, and another section an MFB. This is because there is no interaction or loading or feedback between sections.

For Butterworth filters use Eq. 6.8 with Table 6.1.

$$G(s) = \frac{1}{(s^2 + b_1 s + 1)(s^2 + b_2 s + 1)(\text{etc.})} \tag{6.8}$$

In addition, use the extra factor $(s + 1)$ for odd orders.

TABLE 6.1   BUTTERWORTH TRANSFER FUNCTION COEFFICIENTS

n	2	3	4	5	6	7	8	9	10
b_1	1.4142	1.0000	1.8477	1.6180	1.9319	1.8019	1.9616	1.8794	1.9754
b_2			0.7654	0.6180	1.4142	1.2470	1.6630	1.5321	1.7820
b_3					0.5176	0.4450	1.1111	1.0000	1.4142
b_4							0.3902	0.3473	0.9080
b_5									0.3129

For Chebyshev filters use Eq. 6.9 with Table 6.2.

$$G(s) = \frac{(c^*)(c_1)(c_2)(\text{etc.})}{(s + c^*)(s^2 + b_1 s + c_1)(s^2 + b_2 s + c_2)(\text{etc.})} \tag{6.9}$$

The b and c quadratic coefficients are shown paired in Table 6.2. The c^* binomial coefficients are the unpaired entries. The dc gain will be 1, that is, 0 dB.

TABLE 6.2 CHEBYSHEV TRANSFER FUNCTION COEFFICIENTS

	0.1 dB		0.5 dB		1.0 dB		3.0 dB	
n	b_i	c_i	b_i	c_i	b_i	c_i	b_i	c_i
1		6.5522		2.8628		1.9653		1.0024
2	2.3724	3.3140	1.4256	1.5162	1.0977	1.1025	0.6449	0.7079
3	0.9694	1.6897	0.6265	1.1424	0.4942	0.9942	0.2986	0.8392
		0.9694		0.6265		0.4942		0.2986
4	0.5283	1.3300	0.3507	1.0635	0.2791	0.9865	0.1703	0.9031
	1.2755	0.6229	0.8467	0.3564	0.6737	0.2794	0.4412	0.1960
5	0.3331	1.1949	0.2239	1.0358	0.1789	0.9883	0.1097	0.9360
	0.8720	0.6359	0.5862	0.4768	0.4684	0.4293	0.2873	0.3770
		0.5389		0.3623		0.2895		0.1776
6	0.2294	1.1294	0.1553	1.0230	0.1244	0.9907	0.0765	0.9548
	0.6267	0.6964	0.4243	0.5900	0.3398	0.5577	0.2089	0.5218
	0.8561	0.2634	0.5796	0.1570	0.4641	0.1247	0.2853	0.0888
7	0.1677	1.0924	0.1140	1.0161	0.0941	0.9927	0.0563	0.9665
	0.4698	0.7532	0.3194	0.6769	0.2561	0.6535	0.1577	0.6273
	0.6789	0.3302	0.4616	0.2539	0.3701	0.2305	0.2279	0.2043
		0.3768		0.2562		0.2054		0.1265
8	0.1280	1.0695	0.0872	1.0119	0.0700	0.9941	0.0432	0.9742
	0.3644	0.7989	0.2484	0.7413	0.1994	0.7235	0.1229	0.7036
	0.5454	0.4162	0.3718	0.3587	0.2984	0.3409	0.1839	0.3209
	0.6433	0.1456	0.4386	0.0881	0.3520	0.0703	0.2170	0.0503
9	0.1009	1.0542	0.0689	1.0092	0.0553	0.9952	0.0341	0.9795
	0.2905	0.8344	0.1984	0.7894	0.1593	0.7754	0.0983	0.7597
	0.4450	0.4975	0.3040	0.4525	0.2441	0.4386	0.1506	0.4228
	0.5459	0.2013	0.3729	0.1563	0.2994	0.1424	0.1847	0.1266
		0.2905		0.1984		0.1593		0.0983
10	0.0816	1.0435	0.0558	1.0073	0.0448	0.9961	0.0277	0.9833
	0.2367	0.8619	0.1619	0.8257	0.1301	0.8144	0.0803	0.8017
	0.3687	0.5680	0.2522	0.5318	0.2026	0.5205	0.1250	0.5078
	0.4646	0.2741	0.3178	0.2379	0.2553	0.2266	0.1576	0.2139
	0.5151	0.0925	0.3523	0.0563	0.2830	0.0450	0.1747	0.0323

For elliptic filters use Eq. 6.10 with Table 6.3.

$$G(s) = \frac{(s^2 + a)(s^2 + a_2)(\text{etc.})(c^*)(c_1/a_1)(c_2/a_2)(\text{etc.})}{(s + c^*)(s^2 + b_1 s + c_1)(s^2 + b_2 s + c_2)(\text{etc.})} \tag{6.10}$$

The b and c quadratic coefficients are shown paired in Table 6.3. The c^* binomial coefficients are the unpaired entries. The dc gain will be 1, that is, 0 dB.

For Bessel filters use Eq. 6.11 and Eq. 6.12 with Table 6.4.

$$G(s) = \frac{(c^*)(c_1)(c_2)(\text{etc.})}{(s + c^*)(s^2 + b_1 s + c_1)(s^2 + b_2 s + c_2)(\text{etc.})} \tag{6.11}$$

$$T(\omega) = \frac{b_1(c_1 + \omega^2)}{(c_1 - \omega^2)^2 + b_1^2 \omega^2} + \frac{b_2(c_2 + \omega^2)}{(c_2 - \omega^2)^2 + b_2^2 \omega^2} + \text{etc.} + \frac{c^*}{c^{*2} + \omega^2} \tag{6.12}$$

TABLE 6.3 ELLIPTIC TRANSFER FUNCTION COEFFICIENTS

n	Ω	a_i	0.1 dB A_0	b_i	c_i	(All case A) c^*	1.0 dB A_0	b_i	c_i	(All case A) c^*
2	1.05	1.4387	0.34	0.1508	1.3990		2.82	0.3142	1.1672	0.9478
	1.10	1.7141	0.56	0.2590	1.6259		4.03	0.4583	1.2099	0.8162
	1.20	2.2360	1.08	0.4725	1.9986		6.15	0.6411	1.2358	0.7020
	1.50	3.9271	3.21	1.0682	2.7450		11.20	0.8794	1.2144	0.5910
	2.00	7.4641	7.42	1.6869	3.2141		17.10	0.9989	1.1701	0.5400
3	1.05	1.2054	1.75	0.0897	1.1670	2.8130	8.13	0.1310	1.0388	
	1.10	1.3703	3.37	0.1708	1.2658	2.2408	11.50	0.1953	1.0424	
	1.20	1.6996	6.69	0.3135	1.3941	1.7441	16.20	0.2729	1.0388	
	1.50	2.8060	14.80	0.5793	1.5539	1.2982	25.20	0.3754	1.0237	
	2.00	5.1532	24.00	0.7637	1.6291	1.1168	34.50	0.4340	1.0106	
4	1.05	1.1536	6.40	1.2372	1.6896		15.80	0.8019	0.6849	
		3.3125		0.0752	1.0954			0.0739	1.0107	
	1.10	1.2909	10.70	1.4076	1.4489		20.80	0.7985	0.5670	
		4.3499		0.1335	1.1411			0.1090	1.0097	
	1.20	1.5724	17.10	1.4537	1.1930		27.40	0.7739	0.4638	
		6.2242		0.2169	1.1654			0.1513	1.0062	
	1.50	2.5356	29.10	1.3975	0.8689		39.50	0.7300	0.3643	
		12.0993		0.3473	1.2581			0.2088	0.9988	
	2.00	4.5933	41.40	1.3409	0.7364		51.90	0.7025	0.3192	
		24.2272		0.4325	1.2941			0.2430	0.9932	
5	1.05	1.1334	13.80	0.5338	1.1033	1.1289	24.10	0.3624	0.7698	0.5118
		1.7737		0.0602	1.0578			0.0471	1.0029	
	1.10	1.2593	20.10	0.6594	1.0175	0.9321	30.50	0.4043	0.6885	0.4466
		2.1931		0.0991	1.0827			0.0692	1.0016	
	1.20	1.5211	28.30	0.7583	0.9101	0.7829	38.80	0.4351	0.6071	0.3916
		2.9684		0.1509	1.1117			0.0962	0.9993	
	1.50	2.4255	43.40	0.8341	0.7757	0.6498	53.90	0.4577	0.5171	0.3378
		5.4376		0.2283	1.1497			0.1331	0.9950	
	2.00	4.3650	58.90	0.8582	0.7044	0.5909	69.40	0.4647	0.4719	0.3126
		10.5680		0.2778	1.1718			0.1552	0.9919	

TABLE 6.3 ELLIPTIC TRANSFER FUNCTION COEFFICIENTS

			0.1 dB		(All case A)		1.0 dB		(All case A)	
n	Ω	a_i	A_0	b_i	c_i	c^*	A_0	b_i	c_i	c^*
6	1.05	1.1233	22.10	1.2941	0.8137		32.50	0.6811	0.3337	
		1.4387		0.3030	0.9940			0.1985	0.8388	
		6.5288		0.0468	1.0376			0.0326	1.0005	
	1.10	1.2434	29.70	1.1995	0.6276		40.10	0.6302	0.2668	
		1.1741		0.3889	0.9434			0.2365	0.7789	
		8.8265		0.0739	1.0537			0.0479	0.9994	
	1.20	1.4950	39.60	1.0953	0.4845		50.10	0.5789	0.2133	
		2.2360		0.4709	0.8793			0.2732	0.7146	
		12.9527		0.1092	1.0727			0.0665	0.9977	
	1.50	2.3693	57.80	0.9757	0.3603		68.20	0.5216	0.1646	
		3.9271		0.5548	0.7940			0.3090	0.6383	
		25.8272		0.1608	1.0983			0.0922	0.9949	
	2.00	4.2482	76.40	0.9145	0.3079		86.80	0.4923	0.1433	
		7.4641		0.5933	0.7455			0.3254	0.5976	
		52.3568		0.1934	1.1134			0.1077	0.9930	
7	1.05	1.1175	30.50	0.7248	0.7574	0.6979	40.90	0.4126	0.5071	0.3522
		1.3083		0.1959	0.9690			0.1239	0.8830	
		2.7144		0.0365	1.0263			0.0238	0.9996	
	1.10	1.2341	39.40	0.7452	0.6385	0.5996	49.80	0.4135	0.4287	0.3101
		1.5239		0.2582	0.9333			0.1552	0.8373	
		3.5148		0.0566	1.0377			0.0350	0.9988	
	1.20	1.4799	51.00	0.7424	0.5311	0.5197	61.40	0.4066	0.3596	0.2741
		1.9413		0.3229	0.8883			0.1867	0.7864	
		4.9667		0.0822	1.0513			0.0488	0.9975	
	1.50	2.3365	72.10	0.7189	0.4242	0.4437	82.60	0.3916	0.2911	0.2382
		3.3140		0.3966	0.8267			0.2217	0.7232	
		9.5301		0.1191	1.0697			0.0677	0.9955	
	2.00	4.1800	93.80	0.7003	0.3746	0.4086	104.00	0.3814	0.2592	0.2211
		6.2018		0.4342	0.7906			0.2395	0.6884	
		18.9611		0.1422	1.0807			0.0791	0.9942	

n	ratio									
8	1.05	1.1139	38.90	1.0369	0.4354	49.30	0.5482	0.1915		
		1.2431		0.4473	0.7841		0.2607	0.6404		
		1.9061		0.1373	0.9650		0.0845	0.9118		
		11.0466		0.0291	1.0194		0.0182	0.9994		
	1.10	1.2283	49.00	0.9335	0.3368	59.50	0.4973	0.1520		
		1.4271		0.4928	0.6943		0.2794	0.5638		
		2.3804		0.1847	0.9379		0.1095	0.8759		
		15.1062		0.0444	1.0279		0.0268	0.9987		
	1.20	1.4703	62.30	0.8380	0.2619	72.80	0.4501	0.1210		
		1.7895		0.5234	0.6054		0.2914	0.4910		
		3.2528		0.2361	0.9037		0.1358	0.8351		
		22.3836		0.0639	1.0382		0.0373	0.9977		
	1.50	2.3157	86.50	0.7381	0.1969	96.90	0.4001	0.0930		
		2.9957		0.5420	0.5087		0.2982	0.4138		
		6.0218		0.2982	0.8565		0.1668	0.7830		
		45.0600		0.0917	1.0521		0.0518	0.9962		
	2.00	4.1367	111.00	0.6891	0.1692	122.00	0.3753	0.0808		
		5.5451		0.5455	0.4609		0.2991	0.3760		
		11.7667		0.3315	0.8284		0.1833	0.7536		
		91.7615		0.1090	1.0605		0.0606	0.9952		
9	1.05	1.1114	47.30	0.7105	0.5069	57.70	0.5142	0.3905	0.3426	0.2699
		1.2054		0.2991	0.8199			0.1750	0.7310	
		1.5943		0.1017	0.9669			0.0614	0.9312	
		3.9937		0.0235	1.0149			0.0144	0.9993	
	1.10	1.2243	58.70	0.6835	0.4136	69.20	0.4483	0.3735	0.2822	0.2385
		1.3703		0.3463	0.7478			0.1976	0.6619	
		1.9377		0.1390	0.9453			0.0815	0.9024	
		5.2992		0.0357	1.0216			0.0211	0.9988	
	1.20	1.4638	73.60	0.6472	0.3358	84.10	0.3930	0.3532	0.2315	0.2114
		1.6996		0.3856	0.6725			0.2161	0.5928	
		2.5791		0.1807	0.9182			0.1033	0.8691	
		7.6524		0.0510	1.0296			0.0294	0.9980	
	1.50	2.3016	101.00	0.5994	0.2628	111.00	0.3390	0.3276	0.1834	0.1843
		2.8060		0.4204	0.5857			0.2325	0.5155	
		4.6363		0.2326	0.8807			0.1300	0.8258	
		15.0143		0.0727	1.0405			0.0409	0.9968	
	2.00	4.1074	129.00	0.5727	0.2301	139.00	0.3137	0.3135	0.1616	0.1713
		5.1532		0.4343	0.5407			0.2391	0.4761	
		8.9222		0.2614	0.8582			0.1447	0.8010	
		30.2010		0.0862	1.0471			0.0479	0.9961	

TABLE 6.3 ELLIPTIC TRANSFER FUNCTION COEFFICIENTS

n	Ω	a_i	0.1 dB A_0	b_i	c_i	(All case A) c^*	1.0 dB A_0	b_i	c_i	(All case A) c^*
10	1.05	1.1097	55.70	0.8444	0.2699		66.10	0.4520	0.1235	
		1.1815		0.4927	0.5942			0.2764	0.4825	
		1.4387		0.2126	0.8512			0.1241	0.7926	
		2.5336		0.0786	0.9703			0.0467	0.9449	
		16.8596		0.0194	1.0119			0.0116	0.9994	
	1.10	1.2216	68.40	0.7547	0.2099		78.80	0.4068	0.0978	
		1.3338		0.5025	0.5055			0.2788	0.4110	
		1.7141		0.2556	0.7916			0.1459	0.7321	
		3.2619		0.1087	0.9527			0.0631	0.9213	
		23.1838		0.0293	1.0172			0.0171	0.9989	
	1.20	1.4592	85.00	0.6745	0.1642		95.40	0.3658	0.0777	
		1.6414		0.5005	0.4257			0.2759	0.3473	
		2.2359		0.2950	0.7271			0.1657	0.6693	
		4.5855		0.1429	0.9306			0.0814	0.8937	
		34.5122		0.0416	1.0236			0.0238	0.9983	
	1.50	2.2916	115.00	0.5922	0.1242		126.00	0.3233	0.0596	
		2.6826		0.4873	0.3456			0.2678	0.2837	
		3.9271		0.3347	0.6498			0.1856	0.5962	
		8.7508		0.1867	0.9000			0.1043	0.8574	
		69.7914		0.0591	1.0325			0.0331	0.9973	
	2.00	4.0867	146.00	0.5522	0.1071		157.00	0.3025	0.0518	
		4.8980		0.4768	0.3081			0.2620	0.2537	
		7.4641		0.3528	0.6085			0.1946	0.5578	
		17.3635		0.2114	0.8815			0.1171	0.8363	
		114.4310		0.0698	1.0378			0.0388	0.9967	

TABLE 6.4 BESSEL TRANSFER FUNCTION
COEFFICIENTS AND TIME-DELAY EXPRESSIONS

n	2	3	4	5	6
b_1	3.0000	3.6778	5.7924	6.7039	5.0392
c_1	3.0000	6.4594	9.1401	14.2725	12.7701
b_2			4.2076	4.6493	4.8520
c_2		*2.3222	11.4878	18.1563	33.8609
b_3					11.1089
c_3				*3.6467	24.0398

6.3 DETERMINING CIRCUIT COMPONENTS FOR A MULTIPLE-FEEDBACK LOW-PASS FILTER

Once the order is determined from the specification and the appropriate nomograph, an MFB LP filter can be designed by using the $G(s)$ from Table 6.1 or 6.2. Notice that elliptic filters and inverse Chebyshev filters cannot be designed by MFB, because they have a more complicated numerator (compare Eqs. 3.11 and 6.10 with Eq. 3.8).   A design procedure for MFB LP filters is as follows:

1. Split the $G(s)$ into quadratic factors.   Use the b and c from Table 6.1 or 6.2. The constant in the numerator can also be split into arbitrary factors. H_0 is the dc gain and is equal to $G(0)$.   It can also be split up, using a different H_0 for each section.

2. Calculate $m_0 = \dfrac{b^2}{4c(1 + |H_0|)}.$ (6.13)

3. Pick C_5 somewhat smaller than m_0, using the standard values of 1, 2.2, or 4.7 times the nearest power of 10.

4. Other values are

$$C_2 = 1$$

$$R_4 = \frac{1}{2C_5c}\left[b \pm \sqrt{b^2 - 4\frac{C_5}{C_2}c(1 + |H_0|)}\right]$$

$$R_1 = \frac{R_4}{|H_0|}$$

$$R_3 = \frac{1}{cC_2C_5R_4}$$

5. Now frequency scale the circuit by dividing C_2 and C_5 by k_f (see Section 4.4).

6. Now impedance scale the circuit by multiplying the resistors by k_m and dividing the capacitors by k_m so that the values are in a practical range, and in particular that the capacitor values are 1, 2.2, or 4.7 times some power of 10, since these are most readily available.

Example 6.3

Design a fourth-order 3-dB Chebyshev filter having a dc gain of 6 dB and a cutoff of 1200 Hz, using the MFB active circuit.

Solution Equation 6.9 is set up to give 0 dB at dc. It is

$$G(s) = \frac{0.9031(0.1960)}{(s^2 + 0.1703s + 0.9031)(s^2 + 0.4412s + 0.1960)}$$

Because we want a dc gain of 6 dB, the numerator above must be multiplied by 2, making it $0.9031(0.1960)2$. We now split this $G(s)$ into two separate $G(s)$, arbitrarily splitting the numerator:

$$G_A(s) = \frac{0.9031}{s^2 + 0.1703s + 0.9031} \qquad G_B(s) = \frac{0.392}{s^2 + 0.4412s + 0.196}$$

Thus the dc gains of each section are $|H_{0A}| = 1$, $|H_{0B}| = 2$. Let us design section A first. From Eq. 6.13, $m_0 = 0.004014$. Setting C_5 to a standard value less than m_0, we pick $C_5 = 0.0022$ F. Using the equations in step 4, we have $C_2 = 1.0$ F, $R_4 = 71.67$ Ω, $R_1 = 71.67$ Ω, $R_3 = 7.023$ Ω. The frequency scaling factor, k_f, is $2\pi1200$

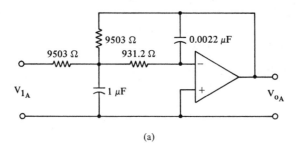

(a)

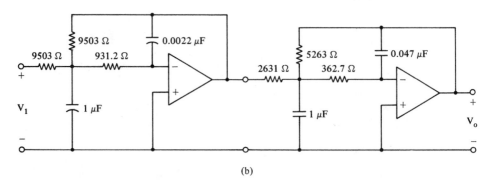

(b)

Figure E6.3

= 7540. This makes C_2 and C_5 = 132.6 μF and 0.2918 μF, respectively. Since the resistors should be larger to be practical (about 100 times as large), we will use a k_m in that range. But since k_m will affect the capacitors also, let us pick a k_m that will bring the capacitors to standard values. Picking k_m = 132.6 will then make C_2 = 1 μF, C = 0.0022 μF, R_1 = R_4 = 9503 Ω, R_3 = 931.2 Ω. Applying these values to Figure 5.4, we have the circuit shown in Figure E6.3a.

We can use the same technique for section B [for $G_B(s)$], but since it is independent of the first section, we are not required to use the same impedance scaling factor as before. However, the frequency scaling factor must be the same, that is, 7540. When this is done and one section tacked onto the other, the complete fourth-order Chebyshev active filter is as shown in Figure E6.3b.

6.4 DETERMINING CIRCUIT COMPONENTS FOR A SALLEN-KEY FILTER

Another method of designing Butterworth and Chebyshev LP filters is the Sallen-Key circuit. The design procedure is as follows:

1. Find the $G(s)$ from Tables 6.1 and 6.2 and split it into quadratic sections as discussed in the MFB procedure. In this case, however, the dc gain of each section and the overall gain must be greater than 1. The dc gain of each section is given by

$$\mu = 1 + r \quad \text{where} \quad r = \frac{R_6}{R_3} = \mu - 1 \tag{6.14}$$

This is a property of Sallen-Key filters. A voltage divider could be attached to the output of the last section to cut down the gain, if desired.

2. Calculate $m_0 = r + b^2/4c$. (6.15)

3. Pick C_2 somewhat smaller than r using the standard values of 1, 2.2, or 4.7 times the nearest power of 10.

4. Other values are

$$C_5 = 1 \quad R_4 = \frac{\sqrt{4c(m_0 - C_2/C_5)} - b}{2c(rC_5 - C_2)} \quad R_1 = \frac{1}{cR_4C_2}$$

5. Now frequency scale the circuit by dividing the capacitors by k_f (see Section 4.4).

6. Now impedance scale the circuit by multiplying the resistors R_1 and R_4 by k_m and dividing the capacitors by k_m, so that the values are in a practical range, and in particular that the capacitors are 1, 2.2, or 4.7 times some power of 10. R_6 and R_3 may be separately scaled, because it is only their ratio that affects anything.

Example 6.4

Repeat Example 6.3 using the Sallen-Key circuit.

Solution Since the overall gain is +6 dB, we can split up the overall $G(s)$ from Example 6.3 into two sections, each of which will have a gain greater than 1, as required by Sallen-Key. Arbitrarily making the dc gain of each section the same by multiplying the c factors by $\sqrt{2}$, we have

$$G_A(s) = \frac{1.2772}{s^2 + 0.1703s + 0.9031} \qquad G_B(s) = \frac{0.27719}{s^2 + 0.4412s + 0.196}$$

and $|H_{0A}| = 1.2772/0.9031 = 1.4142$, $|H_{0B}| = 0.27719/0.196 = 1.4142$. From Eq. 6.14, $r = 0.414$ and $R_6 = 0.414R_3$. Since scaling of these two resistors is arbitrary, let $R_3 = 10$ kΩ and $R_6 = 4.14$ kΩ.

Let us design section A first. From Eq. 6.15, $m_0 = 0.422$. Set C_2 to a standard value less than r, say $C_2 = 0.22$ F. Then, from step 4, $C_5 = 1$ F, $R_4 = 1.952$ Ω, and $R_1 = 2.578$ Ω. Dividing C_5 and C_2 by $k_f = 7540$ makes them 132.6 μF and 29.18 μF, respectively. Resistors R_1 and R_4 must be scaled by roughly 100 or more. Let

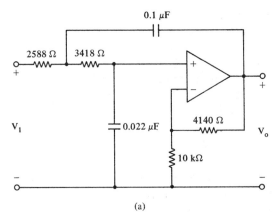

(a)

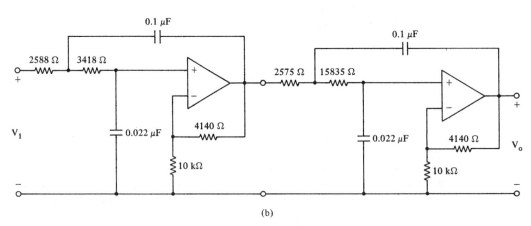

(b)

Figure E6.4

us set $k_m = 1326$. This will bring C_5 and C_2 to the standard value of 0.1 μF and 0.022 μF, respectively. Then R_1 becomes 3418 Ω and R_4 becomes 2588 Ω. Applying these values to the Sallen-Key circuit of Figure 5.6 yields the circuit shown in Figure E6.4a.

Using the same technique for section B and tacking it onto section A will yield the complete fourth-order Chebyshev active filter (Sallen-Key) shown in Figure E6.4b. Notice that this circuit and the one from Example 6.3 have the same $G(s)$, yet their circuitry and component values are different.

6.5 DETERMINING CIRCUIT COMPONENTS FOR A BI-QUAD LOW-PASS FILTER

Active LP Butterworth and Chebyshev filters can also be designed with the bi-quad circuit. As the name suggests, a bi-quad can handle a $G(s)$ which has a quadratic in the numerator as well as in the denominator. Thus it can handle an elliptic LP filter, which the other two circuits could not. The all-purpose bi-quad shown in Figure E5.5e can accommodate the bi-quadratic equation

$$G(s) = -\frac{ds^2 + es + a}{s^2 + bs + c} \qquad \text{if } da > 0 \tag{6.16}$$

Looking at Eq. 6.10 for the elliptic LP, we see that $da > 0$. Also, $e = 0$ (no center term in the numerator). Then a very simple procedure can be followed to find the circuit values.

The values from Table 6.3 are for case A elliptic filters, including the even-order filters. Remember that for passive filters we had to resort to case C for the even-ordered elliptics and our nomographs took this into account. This means that we cannot use the given nomograph when doing even-ordered active elliptic filters. However, Table 6.3 has all the information (Ω, A_{PB}, A_Ω, and n) necessary to find the proper order. This also means that the number of half-ripples equals the order, no matter if it is an odd-order or an even-order filter. So the design procedure for bi-quad LP elliptic filters is:

1. Find $G(s)$ from Table 6.3, using the given specification to find the a, b, and c values pertaining to the required dB ripple, stopband dB, and frequency ratio. Then split this $G(s)$ into bi-quadratic fractions. Multiply one of these fractions by the dc gain, or apply part of the dc gain to each fraction; this will cause the d and a values to change.

2. Design each bi-quadratic section separately. Using Figure E5.5e and the a, b, and c values from Table 6.3, the circuit values for the LP prototype are

$$C_1 = C_2 = 1\text{ F} \qquad R_1 = \frac{1}{bd - e} \qquad R_2 = \frac{1}{b} \qquad R_3 = \frac{1}{c}$$

$$R_4 = \frac{1}{d} \qquad R_5 = \frac{c}{a} \qquad R_7 = R_8 = R_9 = 1$$

3. Now frequency scale the circuit by dividing the two capacitors by k_f (see Section 4.4).

4. Now impedance scale the circuit by multiplying the resistors by k_m and dividing the capacitors by k_m so that the resistors and capacitors are both in a practical range, in particular that the capacitor values are 1, 2.2, or 4.7 times some power of 10.

5. After all bi-quadratic sections are realized, you will have to design a first-order section, using the c^* values from Table 6.3, if it is an odd-order filter. This can be an active or a passive section, frequency scaled, of course.

6. Tack all the sections together and the design is complete. Of course, a bi-quad circuit such as this will have many op-amps.

Example 6.5

Design an LP elliptic filter to meet the specification shown in Figure E6.5a, by using bi-quads.

Solution   From the specification we see that $A_{PB} = 1.0$ dB, $A_\Omega = 50$ dB, and $\Omega = 2.0$. If we consult the nomograph, we see that a fourth-order passive filter would not quite make the specification. However, by consulting Table 6.3, we see that a fourth-order active elliptic filter with a ripple of 1 dB and $\Omega = 2.0$ will have a stopband attenuation of 51.9 dB, thus better than our requirements. So this shows the superiority of the case A filter over the case C filter. Since our specification calls for -1 dB at dc, but Eq. 6.10 will provide 0 dB at dc, we must change the numerator by -1 dB, or 0.891. Using this fact with the a, b, and c values from the table, we get

$$G(s) = \frac{(s^2 + 4.5933)(s^2 + 24.2272)(0.3192/4.5933)(0.9932/24.2272)0.891}{(s^2 + 0.7025s + 0.3192)(s^2 + 0.2430s + 0.9932)}$$

We will split up the group of constants at the end of the numerator by multiplying the first term in parentheses of the numerator by $(0.3192/4.5933)$ $\sqrt{0.891}$ and the second by $(0.9932/24.2272)$ $\sqrt{0.891}$. This will give a split into two $G(s)$, each with -0.5 dB of gain at dc. Thus

$$G_A(s) = \frac{0.0656s^2 + 0.3013}{s^2 + 0.7025s + 0.3192} \qquad G_B(s) = \frac{0.0387s^2 + 0.9375}{s^2 + 0.2430s + 0.9932}$$

Identifying the numbers above with the letters from Eq. 6.16 and doing section A first, using the relations from step 2, we have

$$R_1 = 21.7\ \Omega \qquad R_2 = 1.423\ \Omega \qquad R_3 = 3.133\ \Omega \qquad R_4 = 15.24\ \Omega$$

$$R_5 = 1.059\ \Omega \qquad R_6 = \text{open} \qquad \text{all the other components} = 1$$

Similarly, for section B, we get

$$R_1 = 106.3\ \Omega \qquad R_2 = 4.115\ \Omega \qquad R_3 = 1.007\ \Omega \qquad R_4 = 25.84\ \Omega$$

$$R_5 = 1.059\ \Omega \qquad R_6 = \text{open} \qquad \text{all the other components} = 1$$

Since the capacitors in both sections are 1 and the same frequency scaling must be done ($k_f = 2\pi1000 = 6283$), all capacitors become 159 µF. To get the impedance

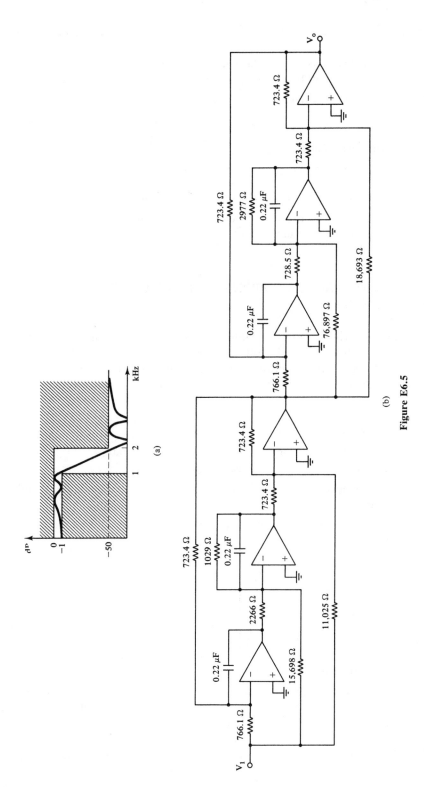

Figure E6.5

level about 1000 times as high and still get standard capacitors, let us set $k_m = 723.4$; this will give us the following values for section A:

$$R_1 = 15,698 \ \Omega \qquad R_2 = 1029 \ \Omega \qquad R_3 = 2266 \ \Omega \qquad R_4 = 11,025 \ \Omega$$

$$R_5 = 766.1 \ \Omega \qquad R_7 = R_8 = R_9 = 723.4 \ \Omega \qquad C_1 = C_2 = 0.22 \ \mu F$$

The values for section B are

$$R_1 = 76,897 \ \Omega \qquad R_2 = 2977 \ \Omega \qquad R_3 = 728.5 \ \Omega \qquad R_4 = 18,693 \ \Omega$$

$$R_5 = 766.1 \ \Omega \qquad R_7 = R_8 = R_9 = 723.4 \ \Omega \qquad C_1 = C_2 = 0.22 \ \mu F$$

Tacking the two sections together yields the final bi-quad fourth-order elliptic LP filter shown in Figure E6.5b.

6.6 DETERMINING CIRCUIT COMPONENTS FOR A STATE-VARIABLE FILTER

Finally, we will consider the state-variable filter as a means of realizing a higher-order LP filter. Actually, this method can be used on HP, BP, and BR filters also. It will involve integrators and can be used to realize almost any $G(s)$, provided that the degree of the numerator is not greater than that of the denominator. The procedure is as follows:

1. Let $V_o(s)/V_1(s)$ equal the given $G(s)$ in expanded form.
2. Cross-multiply so as to get rid of the fraction bars.
3. Divide both sides by the highest power of s.
4. Put the term involving V_o, but no s, on one side of the equation, and everything else on the other side.
5. Do nested factoring of $(-1/s)$ out of the right side of the equation. For instance,

$$3V_1 + \frac{2V_1}{s} - \frac{4V_o}{s^2} - \frac{5V_1}{s^3}$$

$$= 3V_1 + \left(-\frac{1}{s}\right)\left\{-2V_1 + \left(\frac{-1}{s}\right)\left[-4V_o + \left(\frac{-1}{s}\right)(5V_1)\right]\right\}$$

6. Construct a block diagram as an alternating series of $n + 1$ adders and n inverting integrators (where $n =$ highest power of s in any term).
7. Apply the proper amount of V_o (feedback) and V_1 (feedforward) at each adder, in accordance with the factored equation, working from the output end toward the input end. If a negative sign appears in any of these amounts, add an inverter in the circuit.
8. Convert the block diagram to a schematic by using circuit blocks similar to

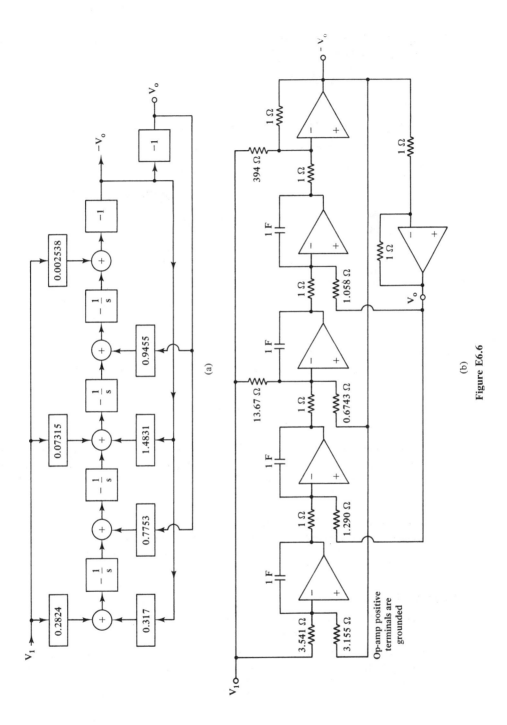

(a)

(b)

Figure E6.6

Figure 5.10, remembering that the resistor values are the reciprocals of the voltage coefficients.

9. Frequency scale and impedance scale the circuit.

Example 6.6

Realize the filter of Example 6.5 by using a state-variable filter.

Solution   Expanding the $G(s)$ from Example 6.5 and cross-multiplying, we have

$$V_o(s^4 + 0.9455s^3 + 1.4831s^2 + 0.7753s + 0.317)$$

$$= V_1(0.002538s^4 + 0.07315s^2 + 0.2824)$$

Dividing by s and putting all terms (except the first) on the right side and arranging in descending powers of s, we get

$$V_o = 0.002538V_1 - \frac{0.9455V_o}{s} + \frac{0.07315V_1}{s^2} - \frac{1.4831V_o}{s^2}$$

$$- \frac{0.7753V_o}{s^3} + \frac{0.2824V_1}{s^4} - \frac{0.317V_o}{s^4}$$

Factoring $(-1/s)$ out of this, in a nested fashion, yields

$$V_o = 0.002538V_1 + \frac{-1}{s}\left\{0.9455V_o + \frac{-1}{s}\left[0.07315V_1 + (-1.4831V_o)\right.\right.$$

$$\left.\left. + \frac{-1}{s}\left(0.7753V_o + \frac{-1}{s}\left(0.2824V_1 + (-0.317V_o)\right)\right)\right]\right\}$$

Setting up five adders and four inverting integrators and an inverter (for the required negative V_o in the equation above) and applying the constants above, we get the block diagram shown in Figure E6.6a.   Finally, we convert this block diagram to the schematic shown in Figure E6.6b.

A final step (not shown) is to frequency scale and impedance scale the circuit. Using the k_m and k_f from Example 6.5 means multiplying all resistors by 723.4 and making all the capacitors 0.22 μF.   Notice that this circuit involves six op-amps as does the bi-quad circuit, but it could be more difficult to adjust, since there is interaction between all the stages due to the various feedback circuits.

Example 6.7

Design a third-order active LP filter to have a nearly constant time delay between 0 and 1 kHz.   How much is the time delay?

Solution   The proper filter is a Bessel filter and its delay is inversely proportional to the cutoff frequency in rad/s.   Thus the delay is the reciprocal of $2\pi 1000$, which is 159 μs.   Using Table 6.4, we have

$$G(s) = \frac{2.3222(6.4594)}{(s + 2.3222)(s^2 + 3.6778s + 6.4594)} = G_A(s)G_B(s)$$

where

$$G_A(s) = \frac{2.3222}{s + 2.3222} \qquad G_B(s) = \frac{6.4594}{s^2 + 3.6778s + 6.4594}$$

Using an MFB circuit for $G_B(s)$ and noting that $|H_0| = 1$, we get $m_0 = 3.6778/[4(6.4594)$ $(1 + 1)] = 0.07117$; so we pick $C_5 = 0.047$ F. Then, using the other formulas for MFB realization, we get $C_2 = 1$ F, $R_4 = 11.544$ Ω, $R_1 = 0.08663$ Ω, $R_3 = 0.2853$ Ω, and $k_f = 6283$. Using this k_f, C_2 becomes 159.2 μF and C_5 becomes 7.480 μF. Finally, to bring up the impedance level, let us make $k_m = 15,920$. This makes $C_2 = 0.01$ μF, $C_5 = 470$ pF, $R_1 = 1379$ Ω, $R_3 = 4542$ Ω, and $R_4 = 183,800$ Ω.

For $G_A(s)$ we could use a simple RC passive network; but since this is an LP section, that would mean a capacitor as a load. Instead, let us use a little op-amp circuit as shown in Figure E6.7a. Because of the low output impedance of op-amps, we could attach any load we want (within reason) without fear of affecting any voltages in the circuit. The gain of this circuit is

$$G_A(s) = -\frac{1/R_6}{s + 1/R_7}$$

Thus $R_6 = R_7 = 0.4306$ Ω.

Frequency scaling by the factor 6283 makes $C_8 = 159.2$ μF. Then impedance scaling by 15,920 yields $C_8 = 0.01$ μF and both resistors become 6855 Ω. Attaching this circuit onto the MFB circuit with the values calculated previously for $G_B(s)$ yields the circuit shown in Figure E6.7b.

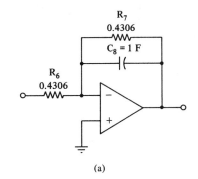

(a)

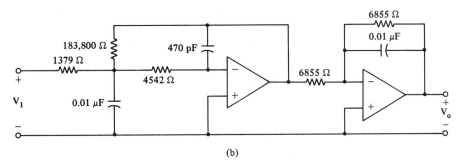

(b)

Figure E6.7

6.7 SYNTHESIZING HIGH-PASS FILTERS BY MODIFYING LOW-PASS ACTIVE FILTERS

High-pass filters can be designed by first designing an LP prototype active filter. The next step, however, is not to use Table 4.7 as we did for the passive case. Although this would work, it would replace all the capacitors with inductors. This is not desirable, because one of the reasons for using active filters was to avoid using inductors. Instead, we will make what is called an *RC–CR* transformation. This means that resistors become capacitors, and vice versa. At the same time we can make the frequency transformation. The rules are:

1. Use any of the previous methods for designing various types of active prototype LP circuits.
2. Change the capacitors to resistors, making $R = 1/C$.
3. Change the resistors to capacitors, making $C = 1/(k_f R)$. Resistors involved in a simple inverter circuit or in the voltage divider (R_3 and R_6) of the Sallen-Key circuit should not be changed to capacitors.
4. Do any desired impedance scaling.
5. In the case of the Sallen-Key circuit, it would probably be possible to devise a procedure to make the capacitors in the HP circuit standard values since there are only two of them. However, this would not be possible for the MFB or the bi-quad circuits, since there are three or more of them now.

Another way of transforming an LP filter to an HP filter is to get an HP transfer function, $G_{HP}(s)$, by using the following formula:

$$G_{HP}(s) = G_{LP}\left(\frac{k_f}{s}\right) \qquad (6.17)$$

This means: substitute k_f/s for s in the LP prototype transfer function. The result will be an HP transfer function having a cutoff at k_f rad/s.

Example 6.8

Find the transfer function for a second-order HP Butterworth filter that cuts off at 200 rad/s by using Eq. 6.17 and confirm that it does cut off at this frequency.

Solution From Table 6.1, $G_{LP}(s) = 1/(s^2 + 1.4142s + 1)$. Now, substituting $200/s$ for s, we have $G_{HP}(s) = 1/[40,000/s^2 + 200(1.4142)/s + 1]$. Multiplying top and bottom by s^2, we get

$$G_{HP}(s) = \frac{s^2}{s^2 + 282.84s + 40,000}$$

It is seen that this has a gain of 1 at the high end ($\omega \gg 200$), and at $\omega = 200$ rad/s, the gain is $-40,000/(-40,000 + j56,568 + 40,000) = -0.707$, which is the defined cutoff for Butterworth filters.

Example 6.9

Design an MFB HP Chebyshev 0.5-dB third-order filter, having a cutoff at 1200 Hz.

Solution   First, design the corresponding LP prototype filter.   From Table 6.2,

$$G_{LP}(s) = \frac{0.6265(1.1424)}{(s^2 + 0.6265s + 1.1424)(s + 0.6265)} = G_A(s)G_B(s)$$

where

$$G_A(s) = \frac{1.1424}{s^2 + 0.6265s + 1.1424} \quad \text{and} \quad G_B(s) = \frac{0.6265}{s + 0.6265}$$

For $G_A(s)$ (the quadratic factor), using the MFB procedure,

$$m_0 = \frac{0.6265}{4(1.1424)(1 + 1)} = 0.04295$$

We must pick C_5 smaller than this value, but it need not be a standard capacitor, since these will later become resistors anyway; so let $C_5 = 0.025$, for example. C_2 is 1, as before.   Plugging in for the other components yields $R_4 = 18.058$, $R_1 = 0.05538$, and $R_3 \leq 1.939$.

The cutoff frequency is $2\pi1200 = 7540$ rad/s.   Now, converting the R to C, and the C to R by the procedure outlined previously, we get

$$C_3 = \frac{1}{7540(1.939)} = 68.4 \ \mu F$$

$$C_1 = \frac{1}{7540(0.05538)} = 2395 \ \mu F$$

$$C_4 = \frac{1}{7540(18.058)} = 7.344 \ \mu F$$

$$R_2 = \frac{1}{1} = 1 \ \Omega \qquad R_5 = \frac{1}{0.025} = 40 \ \Omega$$

Arbitrarily impedance scaling by a factor of 1000 will yield the following reasonable, but nonstandard values: $C_3 = 0.0684 \ \mu F$, $C_1 = 2.395 \ \mu F$, $C_4 = 0.007344 \ \mu F$, $R_2 = 1000 \ \Omega$, and $R_5 = 40 \ k\Omega$.

For the linear factor, we could use a simple RC circuit, since for the HP case, a resistor will be in the "load" position.   Thus if $R = 1$ in the RC section shown in Figure E6.9a, $G_B(s)$ will be

$$G_B(s) = \frac{1/C}{s + 1/C} = \frac{0.6265}{s + 0.6265}$$

or $C_7 = 1.596$ (LP prototype for this section).   Doing the $RC-CR$ transformation,

$$R_7 = \frac{1}{1.596} = 0.6265 \qquad C_6 = \frac{1}{7540(1)} = 132.6 \ \mu F$$

An impedance transformation of 1326 applied to these two components gives us $R = 830.7 \ \Omega$, $C = 0.1 \ \mu F$.   The complete circuit is shown in Figure E6.9b.

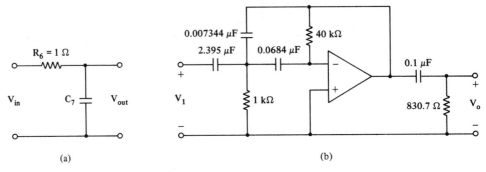

Figure E6.9

6.8 SYNTHESIZING BANDPASS FILTERS BY USING MULTIPLE-FEEDBACK ACTIVE CIRCUITS

There is no easy substitution for getting a BP filter from an LP filter as was done for HP filters with the $RC-CR$ transformation of the preceding section. However, there is an equation similar to Eq. 6.17. It is

$$G_{BP}(s) = G_{LP}\left(\frac{s^2 + \omega_0^2}{Bs}\right) \qquad (6.18)$$

where B is the bandwidth (rad/s) and ω_0 is the center frequency (rad/s). Of course, the output at ω_0 is the same as the output of the prototype at dc, and the output at the frequencies at which the bandwidth is measured is the same as the output at 1 rad/s of the LP filter. When the substitution of Eq. 6.18 is made, the degree of the equation is doubled, meaning that about twice as many components and op-amps may be required. This more complicated $G(s)$ will have to be broken down into quadratic factors as before. Each quadratic factor can be simulated by either a MFB, bi-quad, Sallen-Key, or various other circuits, although the Sallen-Key must be modified by adding a capacitor from R_4 to ground; so we will just look into the other types of circuits.

In the Butterworth and Chebyshev cases, using Eq. 6.18, but letting $B =$ bandwidth divided by center frequency and $\omega_0 = 1$, we will get a transfer function for a "normalized" BP filter, that is, one whose center frequency is 1 rad/s. This can be frequency scaled later. For instance, if

$$G_{LP}(s) = \frac{c}{s^2 + bs + c}$$

then

$$G_{BP}(s) = \frac{cB^2s^2}{s^4 + s^3(bB) + s^2(2 + cB^2) + s(bB) + 1} \qquad (6.19)$$

Doing the same for the elliptic case, using Eq. 6.18 will change the numerator to a quartic:

$$G(s) = \frac{(c/a)[s^4 + (2 + aB^2)s^2 + 1]}{s^4 + bBs^3 + (2 + cB^2)s^2 + bBs + 1} \tag{6.20}$$

If there is a linear term (there will be one if the LP prototype is of odd order), an additional fraction of the following form is produced:

$$G(s) = \frac{c^*Bs}{s^2 + c^*Bs + 1} \tag{6.21}$$

In these equations, a, b, c, and c^* are from Tables 6.1 through 6.4 and B = bandwidth divided by center frequency.

An MFB BP circuit can be constructed to realize the Butterworth or Chebyshev BP filters, but not the elliptic. An all-purpose bi-quad circuit can be constructed to realize any of these filters. If an MFB can be used, we can do an analysis similar to that of Section 5.2, and then we find that the topology for an MFB BP circuit will be op-amp sections such as shown in Figure 6.3. The procedure to find the value of the elements is as follows:

1. Use Eq. 4.2 and the Butterworth or Chebyshev nomograph to find the filter order.
2. Use Table 6.1 or 6.2 to find the prototype LP $G(s)$.
3. Use Eqs. 6.19 and 6.21 to find the $G(s)$ of the BP filter.
4. Use the ROOT FINDER program to find the roots of the denominator. Form factors from these roots.
5. Combine complex-conjugate factors to form quadratic factors. For the numerator above each quadratic, use the nth root of the numerator from step 3 (n = order).
6. Realize an MFB active circuit, similar to Figure 6.3, for each quadratic from

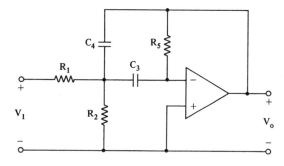

Figure 6.3

step 5 and from Eq. 6.21. Each MFB circuit will have a transfer function of this form:

$$G_{BP}(s) = \frac{e's}{s^2 + b's + c'}$$

7. The values of the components will be

$$R_1 = \frac{1}{e'} \qquad R_2 = \frac{b'R_1}{2c'R_1 - b'} \qquad R_5 = \frac{2}{b'} \qquad C_3 = C_4 = 1$$

8. Frequency scale the capacitors by dividing by $2\pi f_0$.

9. Impedance scale the components above to get reasonable values; the capacitors can be made into standard sizes. It is, of course, not necessary that the same k_m be used for each MFB section.

10. There is a practical difficulty in getting MFB circuits to work properly when $\sqrt{c'}/b'$ is much greater than about 10. In that case, use a bi-quad circuit.

Example 6.10

Design a Chebyshev BP filter to meet the specifications shown in Figure E6.10a by using MFB circuits.

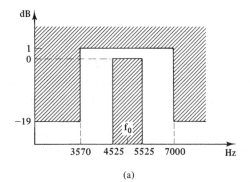

(a)

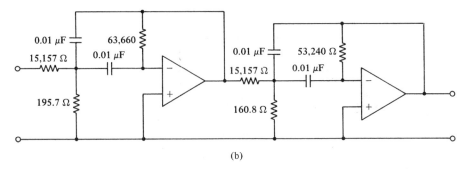

(b)

Figure E6.10

Solution   From the specification we see (using Eq. 4.3) that $f_0 = 5$ kHz.   Using the stopband frequencies in Eq. 4.3, we get the same f_0.   This means that we have a symmetrical specification.   Using Eq. 4.2, we find that $\Omega = 3.43$.   Using this in the Chebyshev nomograph with $A_{PB} = 1$ dB and $A_\Omega = 20$ dB, we find that a second-order prototype LP is required.   From Table 6.2 we get

$$G_{LP}(s) = \frac{1.1025}{s^2 + 1.0977s + 1.1025}$$

Using Eq. 6.19, and letting $B = 1$ kHz/5 kHz $= 0.2$, $b = 1.0977$, and $c = 1.1025$, we get

$$G_{BP}(s) = \frac{1.1025(0.2)^2 s^2}{s^4 + 0.21954s^3 + 2.0441s^2 + 0.21954s + 1}$$

Now, using the ROOT FINDER program, we get

$G(s) =$

$$\frac{(0.21s)^2}{(s + 0.05 + j0.9130)(s + 0.05 - j0.9130)(s + 0.0598 + j1.092)(s + 0.0598 - j1.092)}$$

Combining complex conjugate factors and splitting the numerator, we get

$$G_{BP}(s) = \frac{0.21s}{s^2 + 0.1s + 0.8361} \times \frac{0.21s}{s^2 + 0.1196s + 1.196}$$

Each of the fractions above will now be realized as a separate MFB.   For the first one, using the formulas of steps 6 and 7, we get

$$R_1 = 4.762 \qquad R_2 = 0.0615 \qquad R_5 = 20 \qquad C_3 = C_4 = 1$$

Frequency scaling the capacitors by $k_f = 2\pi 5000$, we get 31.83 μF.   Impedance scaling by $k_m = 3183$, we get

$$R_1 = 15,157 \qquad R_2 = 195.7 \qquad R_5 = 63,660 \qquad C_3 = C_4 = 0.01 \text{ μF}$$

Using the same procedure for the second fraction, we get

$$R_1 = 15,157 \qquad R_2 = 160.8 \qquad R_5 = 53,240 \qquad C_3 = C_4 = 0.01 \text{ μF}$$

Finally, applying these values to two cascaded MFB circuits yields the circuit shown in Figure E6.10b.

6.9 SYNTHESIZING BANDPASS FILTERS BY USING BI-QUAD ACTIVE CIRCUITS

Next, we consider a realization of a BP filter using an all-purpose bi-quad.   The procedure is as follows:

1. Use Eq. 4.2 to find Ω.   Then use the Butterworth or Chebyshev nomographs to get the order.   If it is elliptic, use the information in Table 6.3 to get the order.

2. Use Table 6.1, 6.2, or 6.3 to find the prototype LP $G(s)$.

3. Use Eqs. 6.19 and 6.21 (for Butterworth or Chebyshev), or Eqs. 6.20 and 6.21 (for elliptic), to find the $G(s)$ of the prototype BP filter, by setting B = bandwidth divided by the center frequency. This will give a BP filter with a center frequency at 1 rad/s, and the components can be frequency scaled later. This will avoid any overflow problems in the computer program, and it will avoid huge differences in the sizes of the resistors, when step 6 is performed.

4. Use the ROOT FINDER program to find the roots of the denominator and the numerator; form factors from them.

5. Combine complex-conjugate factors to form quadratic factors in the numerator and denominator (for elliptic filters). Multiply the bi-quadratic fractions thus formed by c/a. If there was a linear term (odd order), or it was a Butterworth or Chebyshev filter, the numerator is just an s term.

6. Realize a bi-quad circuit for each bi-quadratic ratio by using Figure 5.12. The bi-quadratic ratio is of the form

$$G(s) = \frac{d's + e's + a'}{s + b's + c'}$$

then

$$R_1 = \frac{1}{b'd' - e'} \qquad R_2 = \frac{1}{b'} \qquad R_3 = \frac{1}{c'} \qquad R_4 = \frac{1}{d'} \qquad R_5 = \frac{c'}{a'}$$

$$R_6 = \text{open} \qquad \text{and all the other components} = 1$$

7. If the numerator is just an s term (d' and a' are 0), such as Eq. 6.21, it can be realized as an MFB circuit as described previously; or, it can also be realized as a bi-quad, by using Figure E5.5e, and letting

$$R_1 = \frac{1}{e'} \qquad R_2 = \frac{1}{b'} \qquad R_3 = \frac{1}{c'}$$

$$R_7 = R_8 = R_9 = C_1 = C_2 = 1 \qquad R_4 = R_5 = R_6 = \text{open}$$

8. Now divide the capacitors by $k_f = 2\pi f_0$.

9. Impedance scale by dividing all capacitors and multiplying all resistors by k_m, which will cause reasonable values for R and C and standard values for capacitors.

10. Draw the final circuit, by cascading the sections. Practical circuits are possible with Q's as high as 50.

Example 6.11

Realize an active elliptic filter to fit the specifications shown in Figure E6.11a, using the all-purpose bi-quad.

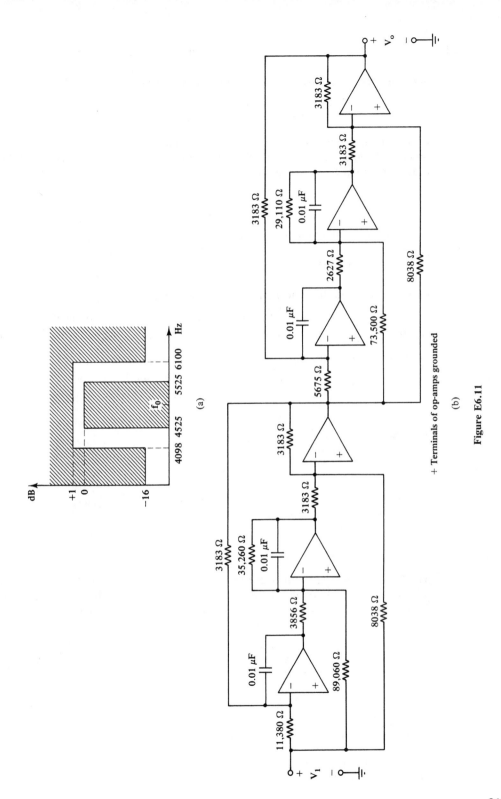

(a)

(b)

+ Terminals of op-amps grounded

Figure E6.11

Solution   As in Example 6.10, f_0 = 5 kHz, BW = 1 kHz, but

$$\Omega = \frac{6100^2 - 5000^2}{6100(1000)} = 2.0$$

Noting that the stopband is 17 dB below the maximum, and consulting Table 6.3, we find that the lowest-order 1.0-dB elliptic filter that has at least a 17-dB stopband gain is a second-order filter. From the table,

$$G_{\text{LP}}(s) = \frac{0.1568(s^2 + 7.4641)}{s^2 + 0.9989s + 1.1701}$$

Using Eq. 6.20, but letting B = 1 kHz/5 kHz = 0.2, we get

$$G_{\text{BP}}(s) = \frac{0.1568(s^4 + 2.2986s^2 + 1)}{s^4 + 0.19962s^3 + 2.0468s^2 + 0.19962s + 1}$$

Factoring this with the ROOT FINDER program, we get

$$\frac{0.1568(s + j0.76343)(s - j0.76343)(s + j1.30987)(s - j1.30987)}{(s + 0.0451 + j0.9074)(s + 0.0451 - j0.9074)(s + 0.0547 + j1.0993)(s + 0.0547 - j1.0993)}$$

Combining factors into quadratics and splitting the 0.1568, we get

$$G(s) = \frac{0.396s^2 + 0.2308}{s^2 + 0.09026s + 0.82541} \times \frac{0.396s^2 + 0.6795}{s^2 + 0.10936s + 1.21152}$$

For the first fraction, d' = 0.396, a' = 0.2308, b' = 0.09026, and c' = 0.82541. Then, using step 6 we have,

$$R_1 = 27.98 \qquad R_2 = 11.079 \qquad R_3 = 1.2115 \qquad R_4 = 2.5253$$

$$R_5 = 3.5763 \qquad R_7 = R_8 = R_9 = C_1 = C_2 = 1$$

For the second fraction, d' = 0.396, a' = 0.6795, b' = 0.10936, and c' = 1.21152. Then, using step 6, we have

$$R_1 = 23.091 \qquad R_2 = 9.1441 \qquad R_3 = 0.8254 \qquad R_4 = 2.5253$$

$$R_5 = 1.7829 \qquad R_7 = R_8 = R_9 = C_1 = C_2 = 1$$

The frequency scaling factor is $2\pi5000$ = 31,416 rad/s. Dividing the capacitors by this, we get 31.83 μF for all capacitors. Now, if we use an impedance scaling factor, k_m, of 3183, all the capacitors become 0.01 μF. All the resistors must be multiplied by 3183 (except R_6, which is open), yielding

	R_1	R_2	R_3	R_4	R_5	R_7	R_8	R_9
First bi-quad	89,060	35,260	3,856	8,038	11,380	3,183	3,183	3,183
Second bi-quad	73,500	29,110	2,627	8,038	5,675	3,183	3,183	3,183

Cascading two all-purpose bi-quads (such as Fig. E5.5e) together and using the components just calculated, we get the circuit shown in Figure E6.11b.

6.10 SYNTHESIZING BAND-REJECT FILTERS BY USING BI-QUAD ACTIVE CIRCUITS

As with BP filters, the $G(s)$ for BR filters can also be found from the $G(s)$ for an LP prototype filter by an appropriate formula as follows:

$$G_{BR}(s) = G_{LP}\left(\frac{Bs}{s^2 + \omega_0^2}\right) \qquad (6.22)$$

The same remarks as for the BP apply, except that neither MFB nor Sallen-Key circuits can be used, since the numerator contains more than one term.   For example, if we take a Chebyshev LP prototype transfer function, which is

$$G_{LP}(s) = \frac{c}{s^2 + bs + c}$$

and perform the substitution according to Eq. 6.22, but normalize the BW and center frequency, we get

$$G_{BR}(s) = \frac{s^4 + 2s^2 + 1}{s^4 + (bB/c)s^3 + (2 + B^2/c)s^2 + (bB/c)s + 1} \qquad (6.23)$$

where B = specification bandwidth divided by specification center frequency, and b and c are from the prototype tables.

Doing the same for the elliptic case, the transfer function becomes the product of the following type of functions:

$$G_{BR}(s) = \frac{s^4 + (2 + B^2/a)s^2 + 1}{s^4 + (bB/c)s^3 + (2 + B^2/c)s^2 + (bB/c)s + 1} \qquad (6.24)$$

For odd-order filters, the LP prototype contains a linear factor.   This will produce the following additional transfer function factor:

$$G_{BR}(s) = \frac{s^2 + 1}{s^2 + (B/c^*)s + 1} \qquad (6.25)$$

The procedure for active BR synthesis is then as follows:

1. Use Eq. 4.2 to find Ω.   Then use the Butterworth or Chebyshev nomographs or, in the case of elliptic, Table 6.3 to get the required order of the filter.
2. Use Table 6.1, 6.2, or 6.3 to find the LP prototype filter transfer function, $G_{LP}(s)$.
3. Use Eq. 6.25, if the filter is of odd order, to get one of the BR fractions.
4. Use Eq. 6.23 to get all the other BR fractions, if it is Butterworth or Chebyshev; use Eq. 6.24, if it is elliptic.   These fractions will have to be factored, using ROOT FINDER.   Then they are re-formed into bi-quadratic fractions.

5. The fractions formed from steps 3 and 4 are all of the form

$$\frac{d's^2 + e's + a'}{s^2 + b's + c'}$$

where $d' = 1$ and $e' = 0$. Each of these fractions produces a bi-quad (see Figure E5.5e) whose normalized values are

$$R_1 = \frac{1}{b'd' - e'} \quad R_2 = \frac{1}{b'} \quad R_3 = \frac{1}{c'} \quad R_4 = \frac{1}{d'}$$

$$R_5 = \frac{c'}{a'} \quad R_6 = \text{open} \quad \text{all the other components} = 1$$

6. Now divide all the capacitors by $k_f = 2\pi f_0$.

7. Impedance scale by dividing all capacitors and multiplying all resistors by k_m so as to get reasonable sizes for the components, and even get standard capacitors.

8. Draw the final circuit by cascading the bi-quad sections together.

Example 6.12

Realize an active Chebyshev filter to fit the specs shown in Figure E6.12a.

Solution   From the specification we see that the center frequency, if calculated as f_0 = $\sqrt{f_1 f_2}$ = 30 kHz or as $\sqrt{f_3 f_4}$ = 30 kHz, yields the same results, indicating a symmetrical specification. Thus, our k_f, to be used later, is $2\pi 30$ kHz = 188,500. The BW = 45 kHz − 20 kHz = 25 kHz. The normalized bandwidth, B, is 25 kHz/ 30 kHz = 0.8333. The value of Ω is

$$\frac{45^2 - 30^2}{45(36 - 25)} = 2.27$$

Using this value and A_{PB} = 1 dB and A_{Ω} = 25 dB, we find from the Chebyshev nomograph that the order must be 3. Taking the values from Table 6.2, we see that

$$G_{LP}(s) = \frac{0.4942}{s + 0.4942} \times \frac{0.9942}{s^2 + 0.4942s + 0.9942}$$

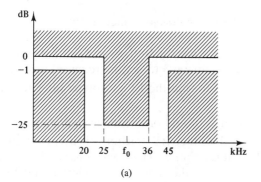

(a)   **Figure E6.12**

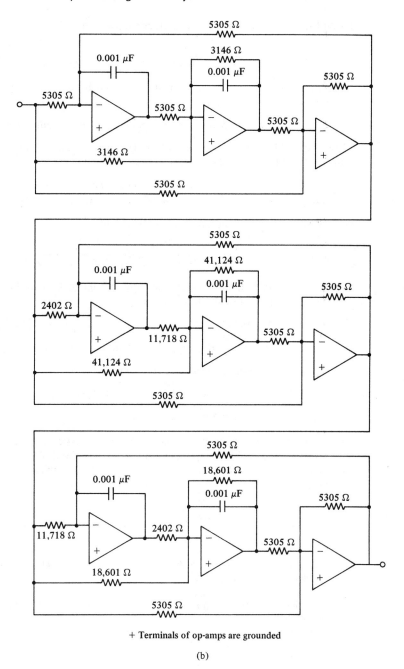

+ Terminals of op-amps are grounded

(b)

Figure E6.12 (Continued)

Using Eq. 6.25 on the first of these fractions, we get

$$G_{BR}(s) = \frac{s^2 + 1}{s^2 + 0.8333s/0.4942 + 1} = \frac{s^2 + 1}{s^2 + 1.686s + 1} \quad \text{(first part)}$$

Since this is a bi-quadratic fraction, it is realized by using the equations of step 5. Since $a' = c' = d' = 1$ and $b' = 1.686$,

$$R_1 = R_2 = 0.593 \qquad R_6 = \text{open} \qquad \text{all other components} = 1$$

Using Eq. 6.23 on the second part of $G_{LP}(s)$ gives a quartic, which must be factored with ROOT FINDER. This yields

$$\frac{(s + j)(s - j)(s + j)(s - j)}{(s + 0.0645 + j0.6697)(s + 0.0645 - j0.6697)(s + 0.1426 + j1.4794)(s + 0.1426 - j1.4794)}$$

When combining into quadratic factors, this yields

$$G_{BR}(s) = \frac{s^2 + 1}{s^2 + 0.129s + 0.4527} \times \frac{s^2 + 1}{s^2 + 0.2852s + 2.206} \quad \text{(second part)}$$

Both portions are bi-quadratic fractions and can be realized as bi-quads, using the equations of step 5. For the first portion,

$$d' = a' = 1 \qquad b' = 0.129 \qquad c' = 0.4527 \qquad \text{so} \quad R_1 = R_2 = 7.752 \qquad R_3 = 2.209$$

$$R_4 = 1 \qquad R_5 = 0.4527 \qquad R_6 = \text{open} \qquad \text{all other components} = 1$$

For the second portion, $d' = a' = 1$, $b' = 0.2852$, and $c' = 2.206$, so

$$R_1 = R_2 = 3.5063 \qquad R_3 = 0.4527 \qquad R_4 = 1$$

$$R_5 = 2.206 \qquad R_6 = \text{open} \qquad \text{others} = 1$$

Using the k_f found previously, all the capacitors become 5.305 µF. Using a k_m of 5305 will give reasonable component values. When the 5.305-µF capacitors are divided by 5305, we get 0.001 µF. The resistors are multiplied by 5305, yielding the following results:

	$R_1 = R_2$	R_3	R_4	R_5	R_7	R_8	R_9
First part	3,146	5,305	5,305	5,305	5,305	5,305	5,305
Second part (first portion)	41,124	11,718	5,305	2,402	5,305	5,305	5,305
Second part (second portion)	18,601	2,402	5,305	11,718	5,305	5,305	5,305

Cascading three all-purpose bi-quads in any order together and using the components just calculated, we get the circuit shown in Figure E6.12b.

6.11 SYNTHESIZING ALL-PASS FILTERS BY USING BI-QUAD ACTIVE CIRCUITS

In Chapter 3 we examined a first-order all-pass filter. Its transfer function was of the form

$$G_{AP}(s) = \frac{s - c^*}{s + c^*} \tag{6.26}$$

It is also possible to achieve a second-order AP filter. Its transfer function would be

$$G_{AP}(s) = \frac{s^2 - bs + c}{s^2 + bs + c} \tag{6.27}$$

By substituting $s = j\omega$ into these expressions and converting to polar form, it is seen that the magnitude of numerator and denominator are always the same; hence the gain is always 1. It does not matter what b and c are, as long as they appear in the numerator and denominator, as shown. The phase has a gradual change as the frequency is varied from zero to infinite frequency. The total change is $360n$ degrees, where n = order. It is possible to get higher than second-order AP functions by combining Eqs. 6.26 and 6.27, or by raising either of these to a higher power. All of these AP filters are perfect filters as far as the gain response is concerned: they are all perfectly flat.

However, an important property of AP filters is the manner in which the phase changes. For many applications it is desirable to have a smooth change; in fact, it is nice to have the phase change directly proportional to the frequency, if possible. But this is exactly what a Bessel filter is good at, although a Bessel's gain does not remain constant. If we make the denominator of Eq. 6.27 a Bessel polynomial and keep the b and c in the numerator the same as the b and c of the denominator, we will have the best of both worlds: an all-pass filter with almost uniform time delay for all the frequencies of interest. In fact, it is better than even the Bessel filter in this regard; but of course it is a more complicated transfer function than the pure Bessel. However, in the active form, it should not be too difficult to synthesize, since we see that Eq. 6.27 is a bi-quadratic, which can be realized in bi-quad form, using the all-purpose filter of Figure E5.5e.

Thus an AP function of order n can be written as

$$G_{AP}(s) = \frac{B_n(-s)}{B_n(s)} \tag{6.28}$$

Because a Bessel function is in the denominator and its conjugate is in the numerator of Eq. 6.28, the time delay is exactly twice what it would be for a pure Bessel filter. Thus if our specification calls for a certain time delay, we only have to find a Bessel function that will produce half of this amount; then the AP given by Eq. 6.28 will give the full delay required.

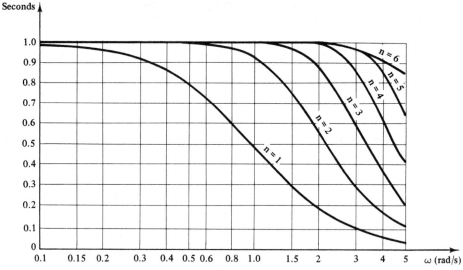

Figure 6.4 Group delay for normalized Bessel filters.

Let us assume that the specification is stated somewhat as follows: An all-pass filter is to have a time delay between 0 and f_0 of at least T_{min} seconds, but no more than T_{max} seconds. The frequency scaling factor will not be $2\pi f_0$. Instead, it will be

$$k_f = \frac{2}{T_{max}} \tag{6.29}$$

This is because of the relation between time delay and frequency, as stated in Section 3.9. To use Figure 6.4, we must normalize our given frequency, f_0, as follows:

$$\omega_{0_N} = \frac{2\pi f_0}{k_f} = \pi f_0 T_{max} \tag{6.30}$$

The normalized value of T_{max} is, of course, 1; but the normalized value of T_{min} is

$$T_{min_N} = \frac{T_{min}}{T_{max}} \tag{6.31}$$

We can now determine the required order from Figure 6.4, by finding the intersection of ω_{0_N} and T_{min_N}. If the intersection lies between two curves, use the higher-order number. $B_n(s)$ can then be found from Table 3.3. However, for our purposes, it is better to get the factored form from Table 6.4. Substitute these factors in the denominator of Eq. 6.28. Replacing s by $-s$ will give us the numerator of Eq. 6.28. Regroup this function into a product of a linear fraction (if it is odd order) times one or more bi-quadratic fractions. These fractions can best be realized as a group of cascaded bi-quad circuits. Thus the procedure can be summarized as follows:

1. Determine, from the specification, the T_{max}, the T_{min}, and the f_0. The frequencies of interest are from dc to f_0. Use Eqs. 6.29, 6.30, and 6.31 to find k_f, ω_{0N}, and T_{min_N}.
2. Find the intersection of T_{min_N} and ω_{0N} on Figure 6.4. If it lies between two order numbers, pick the higher.
3. Using this n in Table 6.4, find $B_n(s)$ in factored form. This is then the denominator of our $G_{AP}(s)$.
4. The numerator of $G_{AP}(s)$ is $B_n(-s)$, found by substituting $-s$ for s in each of the factors of $B_n(s)$.
5. Realize a bi-quad circuit for each linear or bi-quadratic ratio involved. Let all capacitors $= 1$ F. Then the resistors for the bi-quadratic ratios are

$$R_1 = \frac{1}{2b} \qquad R_2 = \frac{1}{b} \qquad R_3 = \frac{1}{c} \qquad R_4 = R_5 = 1$$

$$R_6 = \text{open} \qquad R_7 = R_8 = R_9 = 1$$

The resistors for the linear ratio are

$$R_1 = \frac{1}{2c^*} \qquad R_2 = \frac{1}{c^*} \qquad R_4 = 1 \qquad R_3 = R_5 = R_6 = \text{open}$$

$$R_7 = R_8 = R_9 = 1$$

6. Divide all capacitors by k_f, found in step 1.
7. Divide all capacitors and multiply all resistors by k_m, so as to get reasonable-sized elements and even standard-sized capacitors. Cascade the bi-quad schematics.

Example 6.13

Realize an active AP filter, which between dc and 500 Hz will have a time delay of 1 ± 0.1 ms.

Solution   From the specification we see that $T_{max} = 1.1$ ms, $T_{min} = 0.9$ ms, and $f_0 = 500$ Hz. Using Eqs. 6.29, 6.30, and 6.31, we find $k_f = 1821$, $\omega_{0N} = 1.728$ rad/s, and $T_{min_N} = 0.818$ s. Entering these last two values on Figure 6.4, we find that the required order is 3. From Table 6.4 and Eq. 6.28 we get

$$G_{AP}(s) = \frac{-s + 2.3222}{s + 2.3222} \times \frac{s^2 - 3.6778s + 6.4594}{s^2 + 3.6778s + 6.4594}$$

Make all capacitors 1.0 F. Using the rules of step 5 we have

	R_1	R_2	R_3	R_4	R_5	R_6	$R_7 = R_8 = R_9$
Linear portion	0.2153	0.4306	Open	1.0	Open	Open	1.0
Bi-quadratic portion	0.1360	0.2719	0.1548	1.0	1.0	Open	1.0

Dividing the capacitors by k_f makes all capacitors = 549.1 μF. Then dividing all capacitors and multiplying all resistors by 5491 makes all the capacitors 0.1 μF, and the resistors are

	R_1	R_2	R_3	R_4	R_5	R_6	$R_7 = R_8 = R_9$
Linear portion	1182	2364	Open	5491	Open	Open	5491
Bi-quadratic portion	747	1493	850	5491	5491	Open	5491

Use the values above on two cascaded bi-quad circuits as in Figure E5.5e.

6.12 UNIVERSAL ACTIVE FILTER

We saw in Section 5.4 that a very versatile form of the bi-quad was the all-purpose bi-quad of Figure E5.5e. By choosing the proper element types at the various positions, we could get either an LP, HP, BP, and so on. A "universal" filter is similar to an all-purpose bi-quad, and in the version shown in Figure 6.5, it will give us three different outputs at the same time. That means that we can connect to three different points in the circuit and can obtain three different outputs. One terminal will be a low-pass output (V_{LP}), one terminal will be a bandpass output (V_{BP}), and the third terminal will be either a high-pass, notch (band-reject), or an all-pass output (V_{NHA}). Which one of these three will appear at the third terminal will depend on what combination of the three resistors (R_4, R_5, and R_6) is removed. If none are removed, there is no classification for the type of output that appears at this third terminal.

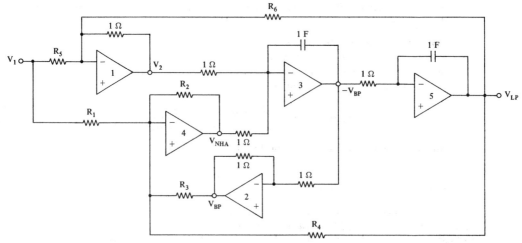

Figure 6.5

Either Butterworth or Chebyshev can be realized.   Elliptic filters can also be realized, but all elliptic responses, whether LP, HP, BP, or BR, look similar to a notch output, so either the V_{NHA} terminal can be used or we can sum the V_{LP} and V_{HP} outputs.

Since the current through the feedback element around each op-amp is equal to the sum of the other currents connected to the input terminal of each op-amp, we can write a KCL equation for each of the five op-amps:

$$\text{Op-amp 1:} \quad V_2 \quad = -\left(\frac{V_1}{R_5} + \frac{V_{\text{LP}}}{R_6}\right) \tag{6.32}$$

$$\text{Op-amp 2:} \quad V_{\text{BP}} \quad = -(-V_{\text{BP}}) \tag{6.33}$$

$$\text{Op-amp 3:} \quad -V_{\text{BP}} = \left(\frac{-1}{s}\right)(V_2 + V_{\text{NHA}}) \tag{6.34}$$

$$\text{Op-amp 4:} \quad V_{\text{NHA}} = R_2\left(\frac{V_1}{R_1} + \frac{V_{\text{BP}}}{R_3} + \frac{V_{\text{LP}}}{R_6}\right) \tag{6.35}$$

$$\text{Op-amp 5:} \quad V_{\text{LP}} \quad = \left(\frac{-1}{s}\right)(-V_{\text{BP}}) = \frac{V_{\text{BP}}}{s} \tag{6.36}$$

Equation 6.33 can be ignored, since this fact is contained on Figure 6.5 explicitly; it is just an inverter.   Thus we have just four equations to solve simultaneously.   The transfer functions at the three terminals relative to the input, V_1, are

$$G_{\text{BP}}(s) = \frac{-s(1/R_5 + R_2/R_1)}{s^2 + s(R_2/R_3) + (1/R_6 + R_2/R_4)} \tag{6.37}$$

$$G_{\text{LP}}(s) = \frac{-(1/R_5 + R_2/R_1)}{s^2 + s(R_2/R_3) + (1/R_6 + R_2/R_4)} \tag{6.38}$$

$$G_{\text{NHA}}(s) = \frac{-[s^2(R_2/R_1) - s(R_2/R_3R_5) + (R_2/R_1R_6 - R_2/R_4R_5)]}{s^2 + s(R_2/R_3) + (1/R_6 + R_2/R_4)} \tag{6.39}$$

We will now remove or possibly change the values of R_1, R_4, R_5, or R_6, so as to make V_{NHA} an AP, HP, BR, or elliptic function.   Table 6.5 shows these values.

If the table says that a resistor is open, it of course means that it is removed from the circuit.   In fact, in the high-pass case, R_5 and R_6 are removed; this means that op-amp 1 is removed, since it has no inputs.   Figure 6.6 shows the proper changes in the four resistors of Figure 6.5 (per Table 6.5); revisions to Eqs. 6.37, 6.38, and 6.39 are also shown with this figure.

In order to use these circuits and equations in a practical problem, we must choose the remaining resistors so as to produce the required Butterworth, Chebyshev, elliptic, or Bessel (LP only) coefficients.   Then we must make a frequency

TABLE 6.5   RESISTOR CHANGES FOR VARIOUS RESPONSES

	Terminal			$V_{HP} + V_{LP}$ (via external op-amp)	
	V_{NHA}				E11 (LP, HP, BP)
Resistor	All-Pass	HP (Bu, Ch)	Some BR	All BR	
R_1	R_2	R_1	R_1	R_1	R_1
R_4	Open	R_4	R_4	R_4	R_4
R_5	$1\,\Omega$	Open	Open	Open	Open
R_6	$1\,\Omega$	Open	$1\,\Omega$	Open	Open

transformation to get the proper cutoff frequency. Finally, we must make an impedance transformation to get reasonable values.

In the case of the AP setup (Figure 6.6a), we see that the numerator and denominator are identical, except for the change in sign of the s term. This proves it is an AP filter (see Eq. 6.27). The coefficient of the s term is R_2/R_3, and although the output will be AP, we can convert it to a Bessel AP (see Section 6.11) if we make sure that all the coefficients are the Bessel coefficients from Table 6.4. However, we see that the constant term in the denominator is 1. Does this mean that we cannot change this value? After all, Bessel filters and Chebyshev filters require values other than 1 at this position for the table prototype values. We can change this value by doing a frequency transformation. This will be shown in Example 6.15. In the case of the BR setup (Figure 6.6c), we see that the constant term in the denominator is always greater than 1. If our prototype filter coefficients call for a number less than 1 in this position, a frequency transformation

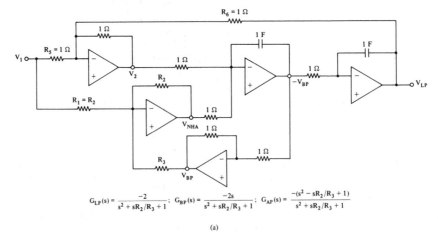

$$G_{LP}(s) = \frac{-2}{s^2 + sR_2/R_3 + 1}; \quad G_{BP}(s) = \frac{-2s}{s^2 + sR_2/R_3 + 1}; \quad G_{AP}(s) = \frac{-(s^2 - sR_2/R_3 + 1)}{s^2 + sR_2/R_3 + 1}$$

(a)

Figure 6.6

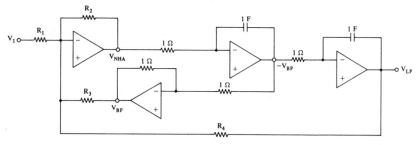

$$G_{LP}(s) = \frac{-R_2/R_1}{s^2 + sR_2/R_3 + R_2/R_4} \; ; \; G_{BP}(s) = \frac{-sR_2/R_1}{s^2 + sR_2/R_3 + R_2/R_4} \; ; \; G_{HP}(s) = \frac{-s^2 R_2/R_1}{s^2 + sR_2/R_3 + R_2/R_4}$$

(b)

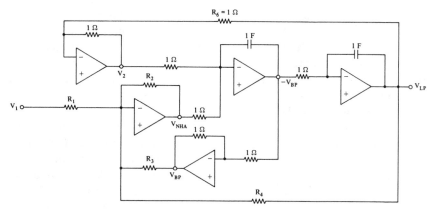

$$G_{LP}(s) = \frac{-R_2/R_1}{s^2 + sR_2/R_3 + (1 + R_2/R_4)} \; ; \; G_{BP}(s) = \frac{-sR_2/R_1}{s^2 + sR_2/R_3 + (1 + R_2/R_4)} \; ; \; G_{BR}(s) = \frac{-R_2/R_1(s^2 + 1)}{s^2 + sR_2/R_3 + (1 + R_2/R_4)}$$

(c)

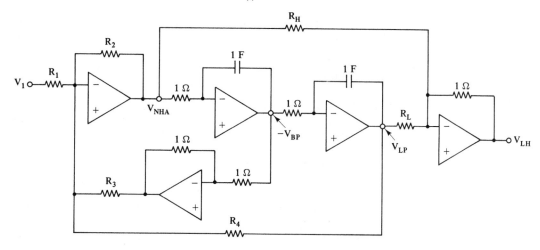

$G_{LP}(s)$, $G_{BP}(s)$, and $G_{HP}(s)$ are the same as for Figure 6.6(b)

$$G_{LH}(s) = \frac{-[s^2 R_2/(R_1 R_H) + R_2(R_1 R_L)]}{s^2 + sR_2/R_3 + R_2/R_4}$$

(d)

Figure 6.6 (Continued)

229

will not work, since that would affect the relationship of the numerator coefficients as well. So for some BR filters and almost all elliptic filters, we must add V_{LP} and V_{HP} of Figure 6.6b, via an external op-amp adder. We can weight V_{LP} and V_{HP} differently to get the proper numerator coefficients.

Since a universal filter can only generate bi-quadratic fractions, we will need to cascade several such universal filters to get the higher-order filters. For instance, if we want to design a BR filter whose LP prototype is a third order, we will get a sixth-order fraction when we apply Eq. 6.24. This will lead to three bi-quadratic fractions, or three universal filters which must be cascaded; that is, the V_{NHA} terminal of the first universal section must be connected to the V_1 terminal of the second, and the V_{NHA} terminal of the second section must be connected to the V_1 of the third section. In the case of filters requiring the external op-amp mentioned in the previous paragraph, we can just put this stage first, then let the inverting op-amp at the start of the next stage do double duty: it can serve as the adder of V_{LP} and V_{HP} of the first section. By this method, an op-amp is saved. The next example will illustrate these points.

Although it is true that an LP function can be achieved with this same circuit, that is true only if we were to reconnect the three sections as follows: V_{LP} of the first section to V_1 of the second section and V_{LP} of the second section to V_1 of the third section; of course in this case, different R values would be required, depending on the LP $G(s)$; and three sections would produce a sixth-order LP filter.

Example 6.14

Repeat Example 6.12 by using the universal filter.

Solution   The synthesis starts off the same as in Example 6.12. From Example 6.12 we see that the transfer function, before frequency transformation, is

$$G_{BR}(s) = \frac{s^2 + 1}{s^2 + 1.686s + 1} \times \frac{s^2 + 1}{s^2 + 0.129s + 0.4527} \times \frac{s^2 + 1}{s^2 + 0.2852s + 2.206}$$

Comparing the first fraction with the $G_{BR}(s)$ of Figure 6.6c, we see that R_4 is an open, $R_2 = R_1 =$ arbitrarily 1 Ω and $R_3 = 0.593$. Similarly, the third fraction can also have $R_2 = R_1 = 1\ \Omega$, but R_3 must be 3.506 Ω and $(1 + R_2/R_4) = 2.206$, making R_4 = 0.829 Ω.

Now, coming back to the second fraction, we see that no value will make $(1 + R_2/R_4) = 0.4527$. Instead, let us use Figure 6.6b and externally add the voltages from terminals V_{LP} and V_{HP}. This is shown in Figure 6.6d. Letting $R_1 = R_2 = R_H = R_L = 1$ will cause the numerator in $G_{LH}(s)$ to be the same as in the second fraction of the problem (except for the minus sign). Then letting $R_3 = 7.752$ and $R_4 = 2.209$ will make the denominator correct. The external op-amp of Figure 6.6d can be eliminated by connecting R_L and R_H of the second stage directly to the first op-amp of the third stage (representing the third fraction), thereby eliminating R_1 of the third stage also.

With frequency transformation, as in Example 6.12, the two capacitors become

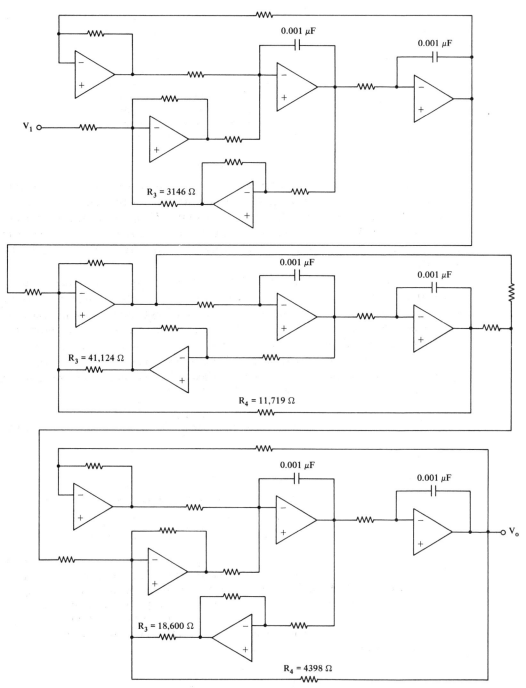

Figure E6.14

5.305 μF (all three sections). Finally, an impedance transformation, using $k_m = 5305$, will make all capacitors $= 0.001$ μF and the resistors will be as follows:

Stage	R_3	R_4	Other R
1	3,146	Open	5,305
2	41,124	11,719	5,305
3	18,600	4,398	5,305

The entire circuit is shown in Figure E6.14.

As we have seen, a universal active filter is quite versatile, but also quite complex. The last example required 14 op-amps for its realization. However, by using switched capacitor technology (discussed in Chapter 7), two op-amps per stage can be eliminated, and the remaining op-amps, capacitors, and most of the resistors can be put on an integrated circuit chip (one IC chip can handle two stages).

If we must realize an odd-order filter (LP or HP), one of the stages must be a first-order stage. For this purpose we will use Figure 6.6b with R_4 open. Notice that the equation for V_{BP} becomes a first-order LP function and V_{HP} becomes a first-order HP function. All BP and BR filters have even-order polynomials in their numerators and denominators, so a first-order stage would never be necessary for these. Finally, for an odd-order AP filter, we could get a first-order stage from Figure 6.6a by letting R_6 be open. This would eliminate the two constant terms in $G_{AP}(s)$ so that (letting $R_1 = R_2 = R_5 = 1$ Ω)

$$G_{AP}(s) = \frac{-(s - 1/R_3)}{s + 1/R_3} \qquad (6.40)$$

Example 6.15

Repeat Example 6.7, but using a universal filter.

Solution The prototype transfer function as worked out in Example 6.7 is

$$G_{LP}(s) = \frac{6.4594}{s^2 + 3.6778s + 6.4594} \times \frac{2.3222}{s + 2.3222}$$

Since this is a third-order LP filter, we can use Figure 6.6a, b, or c for the second-order stage and Figure 6.6b, as discussed above for the first-order stage. Let us use Figure 6.6b for both stages. For the second order-stage, let $R_2 = 1$ Ω. Comparing $G_{LP}(s)$ from Figure 6.6b with the first fraction of our problem, we see that R_3 must be 0.2719 and $R_1 = R_4 = 0.1548$ Ω. For the first-order stage, letting R_4 of $G_{BP}(s)$ of Figure 6.6b be open and letting $R_2 = 1$ Ω, and comparing its simplified version with our first-order fraction, we see that

$$\frac{-1/R_1}{s + 1/R_3} = \frac{2.322}{s + 2.322}$$

so that $R_1 = R_3 = 0.4307$ Ω.

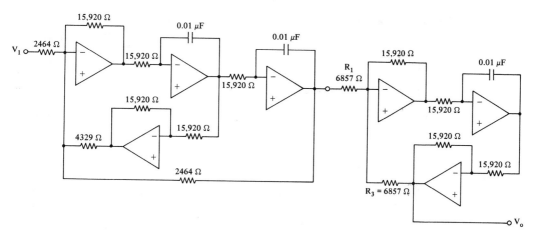

Figure E6.15

Now the circuit must be frequency transformed, using $k_f = 6283$ from Example 6.7. This makes the capacitors = 159.2 μF. Finally, to get reasonable values, use an impedance transformation scaling factor of 15,920. This makes the capacitors all 0.01 μF, and all the resistors are multiplied by 15,920. The final circuit is shown in Figure E6.15.

6.13 COMPUTER TECHNIQUES

Just as we did in the passive case, we can use the computer to prove a synthesis by entering the various components, their values, and their nodes. The output will be the frequency response for gain and phase. These can be plotted as before. However, in this case there is a new component: the op-amp. It is identified in the program by the letter A. It has six parameters in the following order:

1. +IN: node number of the positive input terminal (this is usually the ground node, 0, in most of our problems).

2. −IN: node number of the negative input terminal.

3. +OUT: node number of the output terminal of the op-amp.

4. −OUT: node to which the output terminal is referenced (usually the ground node, 0, in most of our problems).

5. GAIN: theoretically infinite, but the computer can deal only with finite numbers. It should be set to 10,000 for best accuracy, but it may have to be set to 1000 if more than nine op-amps are in the circuit, to avoid overflow in the internal computer calculations.

6. OHMS OUT: output impedance of the op-amp. It is theoretically zero, but an error message will occur if you set it to zero. Actually, it is best to set it to 1 Ω and impedance scale the entire circuit, so that the average impedance

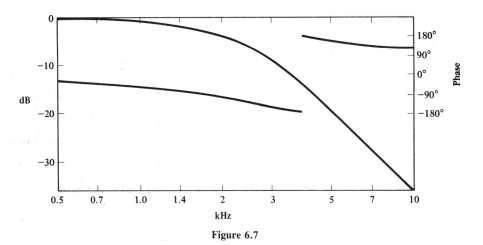

Figure 6.7

is near 1 Ω. This will avoid excessively large or small numbers in the computer during evaluation of the determinants.

If the guidelines above are followed, the program should run and give adequate results. As stated above, the impedance level of the circuit should be scaled back to about 1 Ω. The reason is that at one point in the calculations, the computer multiplies all the admittances together while it is evaluating the nodal determinant. If a number lower than 10^{-38} is generated (in the Apple IIe, for example), an underflow condition is encountered.

Another problem will be running time, which increases approximately proportional to the number of nodes cubed. In some cases you may have to let the computer run overnight. Figure 6.7 shows the output from a 15-node circuit containing seven op-amps. This was run on an Apple IIe using the NETWORK program. The circuit was that of Example 6.15. When properly impedance scaled, it ran properly. Each point required about 2.5 minutes, and it took about 4 hours to calculate the 100 frequencies that were specified. Op-amp gains were specified at 10,000, and OHMS OUT was specified at 1 Ω. The resistors were divided by 10,000 and the capacitors multiplied by 10,000. The error at 500 Hz was less than 0.01 dB, and at 10,000 Hz it was 0.06 dB. Any really large circuit (>20 nodes) would require a much larger mainframe computer.

SUMMARY

In this chapter you learned that you still have to consult the nomographs of Chapter 4 to find the required order of the filter (except for the elliptic case); but from that point on, the procedure is quite different.

A table is consulted so that the proper transfer function for a prototype low-pass filter could be derived. The transfer function thus obtained was in factored form, thus saving the work involved in factoring it with a computer program. These factors were all quadratics and one linear factor (if the order was odd).

This was convenient, since each op-amp section is only capable of synthesizing a quadratic factor. The synthesis for each quadratic factor is carried out by a step-by-step procedure. Procedures are given for multiple-feedback, Sallen-Key, biquad, and state-variable filters for various frequency responses, such as LP, HP, BP, BR, and AP.

The universal filter is also introduced here and it will be used in the next chapter for the switched capacitor realization of active filters.

PROBLEMS

Section 6.1

6.1. $$G(s) = \frac{V_0(s)}{V_1(s)} = \frac{10^{10}}{2s^4 + 3410s^3 + 1.635 \times 10^6 s^2 + 235 \times 10^6 s + 10^{10}}$$

Using ROOT FINDER, break this apart into quadratic factors and realize this function, using MFB synthesis, with equal dc gains in all sections. C_2 must be 2 μF. C_5 must be 1 μF.

Section 6.2

6.2. What are the order, dB ripple, dc gain, and cutoff frequency of an active LP Chebyshev filter, if its transfer function

$$G(s) = \frac{4.053 \times 10^{11}}{s^5 + 323.6s^4 + 147600s^3 + 27.537 \times 10^6 s^2 + 4.368 \times 10^9 s + 2.86665 \times 10^{11}}?$$

(*Hint:* Use ROOT FINDER, group into quadratics, and try to match the ratio of coefficients of like powers with that of table.)

6.3. Find the $G(s)$ of a LP Bessel filter having a time delay of 1 ± 0.1 ms from 0 to 500 Hz, using the minimum order.

Section 6.3

6.4. Realize the $G(s)$ found in Problem 6.3, using MFB synthesis.

Section 6.4

6.5. Realize a Sallen-Key filter with all sections having the same dc gain and all C_5 capacitors 0.01 μF, if the total dc gain is 10 dB, and it is to be a fourth-order LP Butterworth filter having a 3-dB cutoff at 18 kHz.

Section 6.5

6.6. Consider the Bode plot of Figure P6.6.
   (a) Find the $G(s)$.
   (b) Realize it by using only one all-purpose bi-quad such as in Figure E5.5e.   All
          $C = 0.01 \ \mu F$.

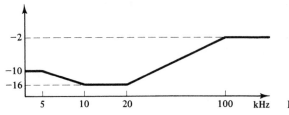

Figure P6.6

Section 6.10

6.7. Realize a BR filter by using the all-purpose bi-quad circuit.   It must meet the spec-
        ifications shown in Figure P6.7.   Use the NETWORK program to prove the circuit
        and attach the computer plot as proof.

6.8. Find the $G(s)$ of an all-pass filter having a time delay of $1 \ \pm \ 0.1$ ms from 0 to 500
        Hz, using the minimum order.

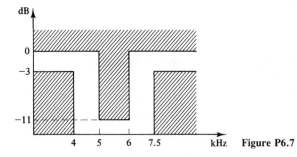

Figure P6.7

Section 6.11

6.9. Realize the $G(s)$ found in Problem 6.8, using all-purpose bi-quad synthesis.

7

introduction to switched capacitors

OBJECTIVES

After completing this chapter, you should be able to:

- Convert an RC passive or active circuit into a switched capacitor circuit.
- Assign proper values to the clock frequency and the capacitors to achieve a circuit that is equivalent to the original circuit.
- Describe some alternative forms of SC circuits and their advantages and disadvantages.
- Design a switched capacitor filter if only the frequency response specifications are given.
- Design a universal active switched capacitor second-order filter using the MF-10 integrated-circuit chip.

INTRODUCTION

Switched capacitor circuits are a relatively new technique that is made practical by the availability of inexpensive IC chips with MOS technology. It permits the construction of active filters without the need of resistors in those portions of the active circuit where resistors usually play a role in the *RC* time constant. Resistors are still used in the other portions of the active circuit, where they have no effect on the frequency response. The chief advantage of this technique is that it is inexpensive, accurate, and takes up very little space. Any active circuit that can be designed by the procedures of Chapter 6 can be converted to its switched capacitor equivalent by the procedures outlined in this chapter. It is even possible to convert a passive *RLC* filter to a switched capacitor filter having only capacitors. This involves a new type of element called a FDNR.

7.1 MOS SWITCH CHARACTERISTICS

A switched capacitor circuit is one in which resistors are replaced by capacitors that are switched in and out of the circuit at a rapid rate. The reasons for doing this and why it works are discussed in the next section. First, we need to find a suitable switch. A mechanical switch will not do, because it does not operate nearly fast enough.

In a bipolar-junction transistor (BJT), the controlling element was the base current. In a metal-oxide-semiconductor field-effect transistor (MOS-FET), the controlling element is the voltage from gate to source (V_{GS}). If this voltage is made large enough, the transistor will let current flow freely between S and D. If V_{GS} is made close to zero, the transistor will cut off (see Figure 7.1). This means that a MOS transistor can behave as a switch. Of course, when this "switch" is in the "on" position, it does not have quite zero ohms resistance. In fact, the resistance may be 10 kΩ. Similarly, when the switch is in the "off" position, the resistance is not infinite; it is about 100 MΩ. However, if this switch is used in a properly impedance-scaled circuit, where the impedance of all the components is much greater than 10 kΩ and much less than 100 MΩ, we can consider for all practical purposes that 10 kΩ is a short circuit and that 100 MΩ is an open circuit. These points are all summarized in Table 7.1.

TABLE 7.1

MOS condition	D-to-S state	Actual D-to-S resistance	Approximate model
$V_{GS} > 2$ V	On	10 kΩ —/\/\/\—	Short o———o
$V_{GS} \approx 0$	Off	100 MΩ —/\/\/\—	Open o———o

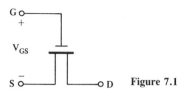

Figure 7.1

7.2 BASIC SWITCHED CAPACITOR CIRCUIT

Consider the circuit shown in Figure 7.2, in which a switch arm is rapidly switched between positions 1 and 2 many times per second. Since $Q_1 = C_1 V_1$ for any capacitor, we see that the capacitor has a charge, $Q_1 = 12 C_1$, as the switch is in movement from 1 to 2. When it reaches 2, the voltage drops down to 5 V almost instantaneously, leaving a charge of $5 C_1$ on the capacitor. Thus it has lost $7 C_1$ of charge, evidently transferred to the 5-V battery. When switched back to position 1, the charge on C_1 suddenly jumps back to $12 C_1$, indicating that C_1 has gained $7 C_1$ of charge, evidently donated by the 12-V battery. Thus this circuit acts as a charging device for the 5-V battery, much the same as Figure 7.3, except that Figure 7.2 transfers charge in a series of current impulses spaced close together in time, and Figure 7.3 transfers charge in a uniform dc current. If the switching frequency is f_c, the average current for Figure 7.2 is $7 C_1 f_c$. In Figure 7.3 the

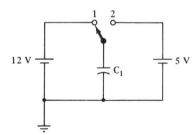

Figure 7.2

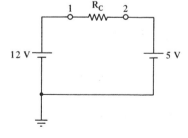

Figure 7.3

current is obviously $7/R_c$. If the currents are the same, these two circuits are roughly equivalent. Equating these two currents, we find that

$$R_c = \frac{1}{C_1 f_c} \tag{7.1}$$

Notice that the number 7 dropped out of this calculation. In fact, Eq. 7.1 holds no matter if sources, opens, shorts, or other capacitors are connected on either side. However, if resistors are connected, Eq. 7.1 can be off by a very large amount, making it unusable. To overcome this problem, any time we convert one resistor to a switched capacitor equivalent, we must convert all other remaining resistors in the circuit to their switched capacitor equivalents also.

The next step is to replace the mechanical switch with the MOS-FET switch mentioned in Section 7.1. However, the mechanical switch was a SPDT switch and the MOS switch was a SPST switch. Thus we must use two MOS switches for each mechanical switch. Figure 7.4 shows the MOS switched capacitor equivalent of Figures 7.2 and 7.3. Voltages are applied to the gate terminals 4 and 5 to control the operation of these switches. These voltages are pulse trains of a bit less than 50% duty cycle and turn these transistors on and off. They are externally generated clock pulses. When the first transistor is conducting, the second is off, and vice versa. It is seen that this will simulate the movement of the switch arm of the mechanical switch.

Thus we are able to change a circuit containing inductors, capacitors, and resistors into one containing capacitors, resistors, and op-amps, thereby eliminating inductors. Now, in this chapter, we get rid of the resistors; all we have left are capacitors, op-amps, MOS transistors, and clock pulses. And the circuitry seems to be more complicated, not simpler. Why would we want to make life more complicated by introducing this new technology?

Let us say that we want to build a small, cheap, accurate filter in mass production. Inductors are bulky and expensive, so we go to active filters, since op-amps are cheap and small. But the capacitors, resistors, and op-amps must all be connected onto a printed circuit board, making for more bulk and expense.

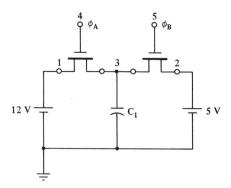

Figure 7.4

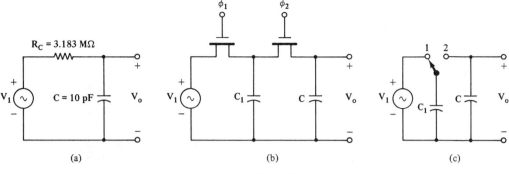

Figure 7.5

We can try to put the capacitors and resistors on the same chip as the op-amp: a special chip. Capacitors on a chip will have very small values; 10 pF is considered a large capacitor in this case. This is okay, because we can impedance scale our design for this. However, when we do this, the resistors are also scaled and become 10 MΩ, if we want a 0- to 4-kHz speech circuit, for instance. It turns out that a 1-MΩ resistor will take up about 10% of the chip, and it will be nonlinear (resistance will change as you increase voltage). Also, the values of C and R can vary by as much as 10%, and these changes are uncorrelated, so that the RC product can be as much as 20% off. In addition, temperature changes will have an additional effect on the RC product.

If we use switched capacitor technology, we get away from the resistor problem and can have almost everything on one chip. This is a relatively new technique. When capacitors that are part of the chip were measured, investigators found that they varied quite a bit from batch to batch and even from one chip to the next, due to uncontrollable variations in the chip-making process. However, they noticed one very interesting thing. If they looked at just one chip at a time and found that one of the capacitors on that chip was 20% higher than it should be, all the other capacitors on that chip were also running about 20% high. This amounts to an accidental impedance scaling of the network on that chip. We have already seen that impedance scaling does not affect the frequency response of a circuit. Since all components (capacitors are practically the only ones) are in error by the same amount, we have a very accurate circuit. Also, a 10-MΩ resistor that takes up 1600 square mils of chip area is replaced by a capacitor that takes up only 3 square mils of chip area, plus the MOSFET switches, which take up negligible area.

For example, the passive LP filter shown in Figure 7.5a has a transfer function

$$G(s) \; = \; \frac{1/sC}{R_c \, + \, 1/sC} \; = \; \frac{1/CR_c}{s \, + \, 1/CR_c} \; = \; \frac{31,420}{s \, + \, 31,420}$$

This is a first-order filter having a 3-dB cutoff at 5 kHz. Figure 7.5b shows the equivalent SC circuit using MOS switches operated at a clock rate of 100 kHz.

Using Eq. 7.1, the capacitor, C_1, which will be equivalent to the 3.183-MΩ resistor, is 3.14 pF. Substituting Eq. 7.1 into the expression for $G(s)$ above gives

$$G(s) = \frac{C_1 f_c / C}{s + C_1 f_c / C}$$

Notice that any variations in the two capacitors will have no effect, as long as both capacitors are in error by the same percentage, since C_1 and C always occur as a ratio in this transfer function. This situation will occur in the transfer function of all *SC* circuits. Since Figure 7.5b is time consuming to draw, we will draw it as Figure 7.5c from now on, with the understanding that the mechanical switch is really a pair of MOS switches.

7.3 ALTERNATIVE FORMS OF A SWITCHED CAPACITOR

Figure 7.6a shows a cross section of a capacitor that is part of an IC chip. Because each chip has a ground plane at the bottom, we notice that there will be substantial capacitance between the ground and the lower plate of the capacitor. This unwanted capacitance is called parasitic capacitance and is labeled C_B in Figure 7.6b. There is very little capacitance between the top plate and ground, because of the shielding effect of the lower plate. Nevertheless, there is a small amount, because of fringing around the edges. This is shown in Figure 7.6b as C_T, and can usually be neglected. However, C_B cannot be neglected, but in the case of Figure 7.2 it can be shorted out by always making certain that the lower plate is the one that is connected to ground.

   Three alternative forms of *SC* implementation are shown in Figure 7.7, involving, in effect, a double-pole double-throw (DPDT) switch. All three require two extra MOS transistors, but they have some interesting properties. They are alternative circuits for Figure 7.5b. The first two cancel out the effect of both the top and lower plate parasitic capacitances, when the output is the input terminal of an inverting op-amp.

   Figure 7.7a shows that each side of C_1 is grounded about half the time. If, in addition, the output is at zero volts potential (as it would be if it is the virtual

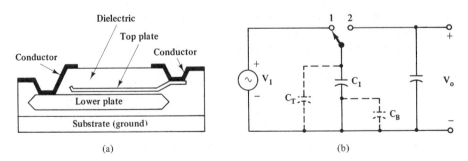

(a)   (b)

Figure 7.6

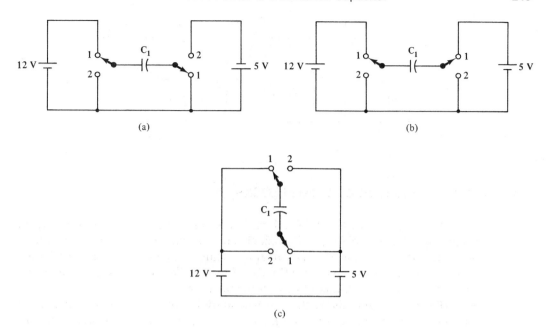

(a)   (b)

(c)

Figure 7.7

ground of an inverting op-amp), this circuit has the effect of preventing any stray capacitance from affecting the circuit. It also has the effect of inverting the voltage, a very useful technique that could avoid the need for an op-amp inverter in some circuits.

Figure 7.7b grounds both sides of the capacitor at the same time and then inserts the capacitor into the circuit. This is also useful if the output is at virtual ground, such as would occur at the input to an inverting op-amp.

Figure 7.7c flips the capacitor end for end twice every cycle. Twice as much charge is transferred to the capacitor each time it changes position and it moves charge to the output twice as often. The net effect is to behave as a resistor one-fourth as large as that given by Eq. 7.1, or, alternatively, it has the same effect as the original R_c, but requiring only one-fourth the clock frequency.

Example 7.1

Realize a high-pass Butterworth first-order SC filter having a cutoff of 10 kHz.

Solution   The normal RC prototype HP circuit consists of a 1-Ω resistor in series with a 1-F capacitor (see Figure E7.1a). Frequency scaling by $2\pi 10,000$ makes the capacitor = 15.9 μF. Assuming a clock frequency of 100 kHz, the 1-Ω resistor becomes a switched capacitor of 10 μF. Finally, impedance scaling is necessary because the capacitors are too large to fit on a chip. Let $k_m = 10^6$. This makes the capacitors as shown in Figure E7.1b. Notice that the replacement technique is to put the two MOS switches in place of the resistor and then put the capacitor from ground to the midpoint between the two MOS switches.

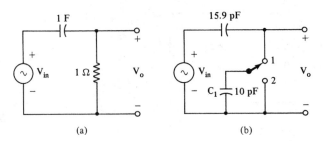

Figure E7.1

7.4 SWITCHED CAPACITOR BUILDING BLOCKS

Switched capacitors are useful in making transconductances, frequency-dependent negative resistors (FDNR), passive ladder filters, negative conductances, gyrators, inductors, amplifiers, integrators, integrator/summers, peak detectors, oscillators, rectifiers, balanced modulators, D/A converters, A/D converters, programmable capacitors, and active bi-quad filters, to name a few.

The first building block that we will study is called a transconductance, or voltage-to-current transducer (VCT). This is a dependent current source; that is, it is a two-port device whose output current is proportional to the input voltage (see Figure 7.8a). As a SC realization (see Figure 7.8b), it consists of a noninverting op-amp (used as a unity-gain buffer) and one switched capacitor. The equivalent transconductance is $g_m = C_m/T = f_c C_m$. Notice that C_m charges to V_{in}, without loading V_{in}, because of the buffering action of the voltage follower. During phase 2, C_m discharges completely through any connected load, since the unity-gain buffer forces the voltage on both sides of C_m to be identical, thus discharging C_m. Because the charge cannot go into the buffer's input (infinite input impedance), the charge must go through the load; hence C_m acts as a current source.

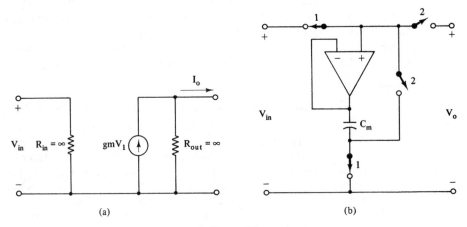

Figure 7.8

The VCT, itself a building block, can be used to create another building block, the FDNR (see Section 5.1).   Figure 7.9 shows such a circuit, although the VCT portion is slightly different from that of Figure 7.8.   The object in this circuit is to find its input impedance:

$$Z_{in} = \frac{V_{in}}{I_{in}} \tag{7.2}$$

From Figure 7.9, we see that

$$V_1 = \frac{I_{in}}{sC_1} \tag{7.3}$$

During phase 1, the op-amp charges C_m to V_1, without disturbing the charge on C_1.   During phase 2, C_m is connected across the buffer, causing C_m to discharge completely, sending all its charge to C_2.   Thus

$$V_2 = \frac{I_m}{sC_2} = \frac{V_1 g_m}{sC_2} = \frac{V_1 f_c C_m}{sC_2} \tag{7.4}$$

But because of the unity gain of the buffer, its output is also V_2, as is the input to the two terminals of the inverting op-amp. Therefore:

$$V_{in} = \frac{V_1 f_c C_m}{sC_2} \tag{7.5}$$

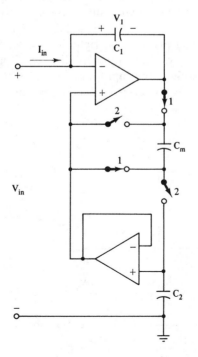

Figure 7.9

Substituting Eq. 7.3 into Eq. 7.5, we get

$$V_{in} = \frac{I_{in}f_cC_m}{s^2C_1C_2} \tag{7.6}$$

Thus

$$Z_{in} = \frac{1}{s^2(C_1C_2/f_cC_m)} = \frac{1}{s^2D} \quad \text{where} \quad D = \frac{C_1C_2}{f_cC_m} \tag{7.7}$$

The letter D is called the FDNR impedance conversion constant. Of course, an equation similar to Eq. 7.7 could have been obtained simply by cascading two op-amp integrators, but this would have been a V_o/V_1 transfer function, not a driving-point impedance. The reason we prefer the driving-point impedance is that now Figure 7.9 can be considered as a one-port; that is, it is a new element, which can be used with capacitors and resistors (SC) to build filters that have topologies very similar to passive RLC filter circuits.

Another building block is the SC amplifier. Although it might seem possible to build an amplifier with an inverting op-amp and two capacitors, there would be the practical problem of having no path for dc currents at the input terminals of the op-amp. So we must use two resistors instead of two capacitors. For the SC realization, however, we can use two switched capacitors. We might want to add a buffer at the input so as not to load the previous stage. Since this buffer is connected to the voltage sampling capacitor only during phase 1, it is possible to have the same buffer do the amplifying during phase 2, so that only one op-amp is used. This interesting circuit is shown in Figure 7.10. This is an inverting amplifier. Although during phase 1 the right side of C_m has a positive voltage, when phase 2 comes into play, C_m completely discharges into C_R, making the top plate of C_R negative, as well as the input and output of the buffer. This means that both plates of C_m are negative, as well as the output terminal, V_o. By rear-

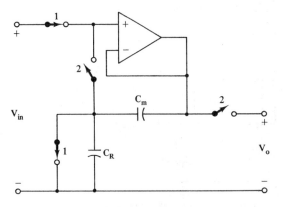

Figure 7.10

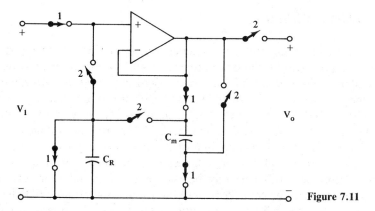

Figure 7.11

ranging the parts as shown in Figure 7.11, it is possible to get a noninverting amplifier. The gain of this amplifier will be C_m/C_R.

Another building block is the SC integrator. If we replace the SC feedback resistor, C_R, of either Figure 7.10 or 7.11 with a normal capacitor, we will have an integrator. The first will provide integration with inversion of sign; the second, just integration. To obtain a "normal" capacitor, all we need to do is to remove the phase 1 switch at the left of C_R in the figures.

The concept of negative resistance (or negative conductance) was discussed in Section 5.1. Figure 7.12 shows an SC realization of a negative conductance. Because of the buffer, during phase 1 no current is taken from the source, but capacitor C_m charges up to the input voltage, V_1. During phase 2, C_m is across the buffer, so it must discharge completely and instantaneously. Since the output of the buffer can supply the discharge current but the input of the buffer cannot accept any current, this current must be sent back to the input source.

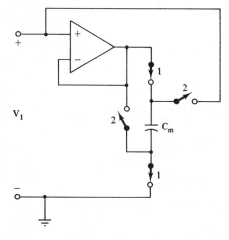

Figure 7.12

7.5 SWITCHED CAPACITOR FILTERS

There are two approaches for realizing switched capacitor filters. The first approach is to design a passive filter and then replace each component by its switched capacitor equivalent. The second approach is to design an active filter as in Chapter 6 and then replace the resistors by their switched capacitor equivalents.

Consider an *LC* ladder filter such as in Example 4.11. If we write in the *s*-domain impedances, the circuit is as shown in Figure 7.13a, and the voltage transfer ratio, $G(s)$, will depend on these *s*-domain impedances. For any circuit, the $G(s)$ will not change, if we divide all the impedances by *s*, as shown in Figure 7.13b. However, the components will certainly be different. The inductors are now resistors; the resistors are now capacitors; and the capacitors are now FDNRs, as shown in Figure 7.13c. Such a transformation of the types of elements is called a **Bruton transformation**.

Notice that Figure 7.13c is a circuit that can be converted to an SC equivalent circuit. The new resistors will now be switched capacitors and the FDNRs will be replaced by circuits such as Figure 7.9. It is also necessary to add a buffer on the input and output ends of this filter to avoid loading or the wrong type of load (the original terminations from Chapter 4 were resistive; the new terminations are capacitive).

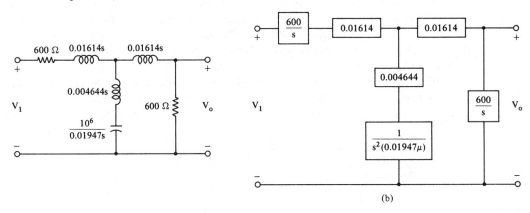

(b)

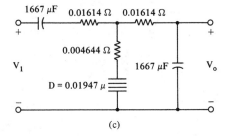

(c)

Figure 7.13

Example 7.2

Convert the circuit shown in Figure 7.13a to its SC equivalent after performing a Bruton transformation.

Solution The Bruton transformation is shown in Figure 7.13c. However, the capacitors are much too high and the resistors too low, so an impedance transformation must be done. In Section 7.2 it was stated that 10 pF is considered large; so let k_m = 2×10^8. This will make the resistors and capacitors as shown in Figure E7.2a. Since the D value of the FDNR appears in the denominator of the impedance expression for it, the new D is 9.73×10^{-17}, as shown in Figure E7.2a. To find the equivalent capacitors to replace the resistors, solve Eq. 7.1, by assuming a clock frequency of

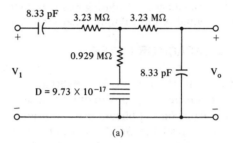

(a)

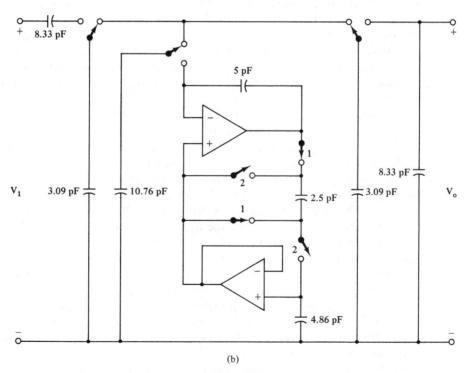

(b)

Figure E7.2

100 kHz. This will give 3.09 pF, which is a reasonable size. If this had turned out larger than 10 pF, we could have picked a smaller clock frequency, or perhaps a larger k_m in the previous step. In place of the FDNR of Figure E7.2a, we put the building block shown in Figure 7.9. Solving Eq. 7.7, we get

$$\frac{C_1 C_2}{C_m} = 9.73 \times 10^{-17} \, (100 \text{ kHz}) = 9.73 \times 10^{-12}$$

If we make all three capacitors 10 pF, this will be approximately satisfied. To satisfy it exactly, and to use smaller capacitors, we see that we can let $C_m = 2.5$ pF, $C_1 = 5$ pF, and $C_2 = 4.86$ pF. The SC equivalent circuit is shown in Figure E7.2b. It is important that at the node where all three resistors are connected, the switches should connect them simultaneously (during phase 2). Of course, buffers may be added at the input or output of this LP third-order filter, if desired.

7.6 MF-10 SWITCHED CAPACITOR FILTER

The most convenient active filter to convert to SC form is the universal filter of Section 6.12. It is especially convenient because the portion of the circuit which does not change from one application to another has already been put on an IC chip and is readily available on the market. It is called the MF-10 Universal Monolithic Dual Switched Capacitor Filter, by National Semiconductor. Although there may be other brands on the market that perform a similar function, the techniques would probably be similar. There are many different modes of operation of this chip, but we will be studying the ones that will fit in with the modes we discussed in Section 6.12. Although the spec sheet talks about Q and 3-dB center frequency of each quadratic factor, we will restrict our synthesis to picking resistances which will make the equations of Figure 6.6 equal to the factors of the $G(s)$ that we are trying to simulate.

By studying Table 6.5 and Figure 6.5, we see that the capacitors are 1 F and most of the resistors are 1 Ω (or are open). Thus the shape of the response depends on which of the circuits in Figure 6.6 are used and on the values of resistors $R_1, R_2, R_3,$ and R_4, as is evident in the equations accompanying Figure 6.6, although Figure 6.6d may require two extra resistors and an external op-amp. These are the only resistors that must be outside the chip.

The following discussion shows how the MF-10 simplifies the realization of Figure 6.6 considerably. Two of the op-amps (1 and 2 of Figure 6.5) and their resistors become unnecessary in the SC version. This is because their major function is to invert a voltage. We know that inversion can be realized in the SC form by using the alternative SC form of Figure 7.6a. If op-amps 3 and 5 were noninverting, the sign of V_{LP} would be unaffected, but the output of op-amp 3 would be $+V_{BP}$, meaning that inverter op-amp 2 can be omitted. All that is required to make op-amps 3 and 5 noninverting is to use the alternative form of Figure 7.6a for the resistors feeding those op-amps. However, it is not necessary to do this for the 1-Ω resistor between op-amps 1 and 3: we just connect R_5 and/or R_6 directly to op-amp 3, thus eliminating op-amp 1.

Since both R_5 and R_6 always have the same value (1 Ω), except when they are opens, they can be part of the chip. We just need some way of connecting them or disconnecting them from the circuit. For R_6, this is done with an on-chip gate, operated by tying the $S_{A/B}$ pin to dc supply or to ground. Since one end of R_5 is connected to the input voltage, it can be disconnected by just not connecting V_1 to it.

So we see that our circuit will have op-amps, capacitors, and SC resistors on the chip, as well as "normal" resistors attached outside the chip. In Section 7.2 it was stated that if some of the resistors are converted to their SC equivalent, all the resistors connected to it must be converted. However, because the output voltage of the op-amp is not loaded down by components following it, we can use normal resistors at one op-amp and SC resistors at the next op-amp, without difficulty. We will be using SC resistors at the op-amps that perform integration (3 and 5), and normal resistors at the adder (op-amp 4). Examining Figure 6.5, we see that the external resistors (R_1, R_2, R_3, and R_4) are all connected to op-amp 4.

The next step involves frequency scaling and impedance scaling. Because of the independence of the op-amps, we can have a different impedance scale factor for each op-amp, if we like. The size of the capacitors and switched capacitors in the chip is quite small (on the order of 10 pF or less), and at the typical combination of audio frequencies and clock frequencies, they have an impedance near 10^6 Ω; whereas the external resistors can be specified by the user. However, if resistors < 1000 Ω are used, and the signal is in the neighborhood of 2 or 3 V, the op-amps could saturate, since they can only source about 3 mA. These resistors are usually impedance scaled to the 10^4 range.

The integrators are the only portions that are frequency scaled. However, instead of changing the capacitors, the SC resistors are changed, since this will have the same effect and is easily controlled by the clock frequency. Since the clock is an extra external component, it can be custom designed for our purpose. It can be either an adjustable clock (555 timer circuit) or a precision fixed-frequency crystal-controlled clock. Examination of the equations involved with Figure 6.6b, c, and d will show that the cutoff frequency of each quadratic can be controlled to some extent also by the ratio R_2/R_4. The effect of the clock frequency is to frequency scale the circuit by $2\pi f_{\mathrm{clk}}/100$ or $2\pi f_{\mathrm{clk}}/50$, depending whether pin 12 is tied at midsupply (0 if two supplies are used) or if it is tied high. Since all the prototype equations we have been working with in this book have a cutoff frequency or center frequency of 1 rad/s, this has the effect of shifting these frequencies to $f_{\mathrm{clk}}/100$ or $f_{\mathrm{clk}}/50$, expressed now in hertz (not rad/s). The range of clock frequencies that will be accepted by the MF-10 is anywhere from 100 kHz to 1.5 MHz, a 15:1 range. When combined with the 50 or 100 factor, this will give us a range of 30:1 for the cutoff frequency for each factor of $G(s)$; that is, it can be from 1 kHz to 30 kHz. If we vary the R_2/R_4 ratio between 1:10 and 10:1 in Figure 6.6b or d, we can get an even larger range: 300 Hz to 100 kHz, large enough to cover most of the audio range.

The MF-10 chip has 20 pins and contains two completely independent second-

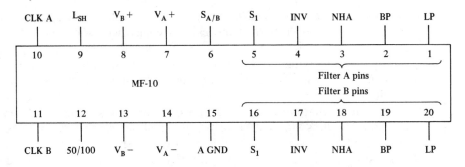

Figure 7.14

order SC filters that can be cascaded to make a third- or fourth-order filter. A different clock, but only one clock factor (100 or 50), can be used for the two filters on each chip. We have seen that the R_2/R_4 ratio can be different for each section: this should give us enough flexibility to design most of our filters. By using more than one chip, we can get filters even higher than fourth order, and we can even apply different external clocks to the additional chips.

The pin-out for the MF-10 chip is shown in Figure 7.14. INV is the pin where V_1 is connected. S_1 is an additional pin to which V_1 must be connected when R_5 is used (Figure 6.6a). The L_{SH} pin is normally tied to ground. The other pins are fairly obvious or were discussed before. For a more complete description, see the manufacturer's specification sheet.

Example 7.3

Design a third-order LP Butterworth filter with a cutoff frequency of 500 Hz and a dc gain of 5 dB.

Solution   From Table 6.1 we see that the LP prototype transfer function is

$$G(s) = \frac{1.333}{s^2 + s + 1} \times \frac{1.333}{s + 1}$$

The numerator constants were picked to provide the 5-dB dc gain. Looking at the four circuits of Figure 6.6, we see that Figure 6.6a will not do, since the dc gain is 2. We could pick any of the other 3 circuits. Arbitrarily picking Figure 6.6b, since it seems the simplest, we see that we can get our quadratic factor by letting $R_1 = 0.75$ Ω and $R_2 = R_3 = R_4 = 1 \Omega$. But since we need a relatively small frequency scaling factor, we will let $f_{clk} = 100$ kHz and the clock factor = 100. This will bring the cutoff to as low as we can get it: 1 kHz, which is still too high. But as we noted before, we can adjust R_2/R_4 to frequency scale it downward slightly. We need a factor of 2. So we form

$$G(2s) = \frac{0.3333}{s^2 + 0.5s + 0.25} \times \frac{0.6666}{s + 0.5}$$

Now when we apply our clock frequency, we get exactly 500 Hz cutoff. If we let $R_2 = 1$, then R_4 must be 4; $R_3 = 2$, and $R_1 = 3$; and the output is at V_{LP}. This is connected to the input (V_1) of the second section. For this second section (the first-

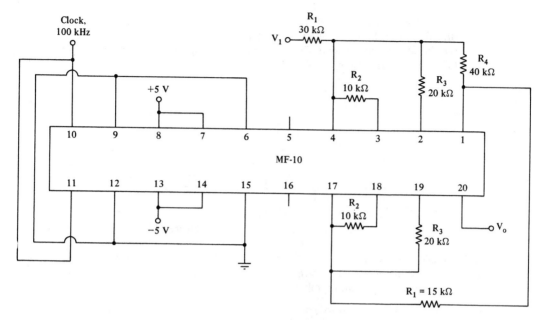

Figure E7.3

order fraction), $R_2 = 1$, $R_3 = 2$, $R_1 = 1.5$, and $R_4 =$ open; and the output is at V_{BP}. Finally, we impedance scale by letting $k_m = 10^4$, which gives us the circuit shown in Figure E7.3.

SUMMARY

Capacitors that are switched from one portion of a circuit to another at a rapid rate by electronic switches (MOSFETs) serve the same function as a resistor would, except that the flow of current is somewhat choppy. If the switching is done at a high frequency, this choppiness is not objectionable or noticeable. Also, by combining switched capacitors with op-amps, not only can the filters of Chapter 6 be realized, but also many new circuits can be realized that are impractical by any other means. The chopped character of the switched capacitor circuit is somewhat of a transition between analog circuits of previous chapters and the digital circuits of Chapter 8. For instance, the MF-10 integrated-circuit chip can be used to build switched capacitor analog filters or it can be used to process digital signals.

PROBLEMS

Sections 7.1, 7.2 and 7.3

7.1. An alternative SC form, somewhat different from that of Figure 7.7c, is proposed as shown in Figure P7.1.

(a) Derive an equation for the current.
(b) What will the final voltage on C_2 be?
(c) What is the equivalent Thévenin circuit, for Figure P7.1, if f_{clk} = 100 kHz?
(d) Find an expression for $V_o(s)/V_1(s)$.

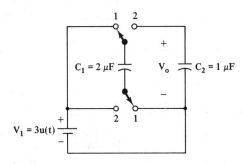

Figure P7.1

7.2. Consider Figure P7.2.
(a) Find the equivalent active circuit.
(b) Solve for the output voltage, $v_o(t)$.

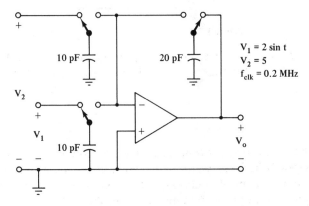

Figure P7.2

7.3. Consider Figure P7.3.
(a) Find $G(s) = V_o(s)/V_1(s)$ if f_{clk} = 468,350 Hz.
(b) Sketch the asymptotes of the Bode plot.

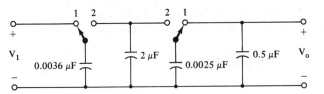

Figure P7.3

7.4. Consider the circuit in Figure P7.4.
(a) Draw the equivalent SC circuit using the alternative form of Figure 7.7a.
(b) Show all component values if f_{clk} = 500 kHz.

(c) Because Figure 7.7a changes the sign of the voltage, does the output of the circuit have the same sign as it would if Figure 7.6b had been used, or would it be opposite? Why?

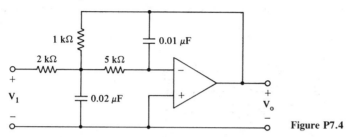

Figure P7.4

Sections 7.4 and 7.5

7.5. Using a Bruton transformation, redraw Figure P7.5, using its SC equivalent, if $f_{clk} =$ 200 kHz. Show all values.

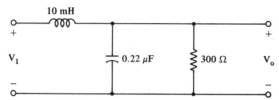

Figure P7.5

Section 7.6

7.6. Find the transfer function, $G(s)$, for the MF-10 circuit shown in Figure P7.6.

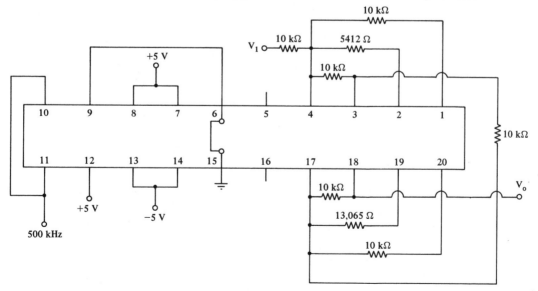

Figure P7.6

7.7. Consider the specifications shown in Figure P7.7.
   (a) Find the $G(s)$ for the corresponding LP prototype filter (Chebyshev).
   (b) Find the $G(s)$ for the normalized BP filter.
   (c) Synthesize a filter that will meet these specifications using the MF-10 chip.

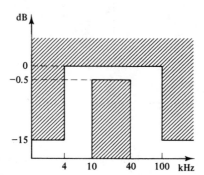

Figure P7.7

8

introduction to digital filters

OBJECTIVES

After completing this chapter, you should be able to:

- Define the z-transform.
- Find the transfer function from a digital block diagram.
- Construct a block diagram from a given $G(z)$.
- Apply the z-transform theorems.
- Apply partial fraction expansion techniques in conjunction with z-transform table pairs to obtain the time function corresponding to a given $F(z)$.
- Determine if a digital circuit is stable or unstable.
- Convert a $G(s)$ to a $G(z)$ by the frequency transformation method or by the time-sampling method; and in such a way as to preserve accuracy over a given frequency range.
- Frequency scale a low-pass digital circuit and convert it to a high-pass digital circuit.

INTRODUCTION

Up until now we have been considering only signals that are continuous functions of time. For instance, if $v(t) = \sin t$, then v at one instant of time will most likely be different from v at some other instant of time, no matter how close the two times are to each other. This is representative of most of what we have been studying. Nowadays, however, discrete signals are quite common. These signals have values only at certain instants of time (at periodic intervals). For instance, in taking data in the lab and plotting them on graph paper, we are plotting discrete data. The data points may be very accurate, but the smooth curve that we use to connect these points is just a presumption on our part, which can be quite inaccurate if the space between points is large. Digital data are discrete data that are rounded off to the nearest bit as defined by the digital base we are using. Thus the distinction between digital and discrete is somewhat arbitrary, and we will use the term "digital" from now on.

8.1 DIGITAL FILTER ELEMENTS

Since a filter circuit is just a device that takes an input signal, processes or modifies it based on the element values and their arrangement in the circuit, and gives this modification as its output, it is no different from a computer program that is fed some data and is expected to give an output based on the equations contained in the program. The only difference is that the computer is working with discrete data. So to have the digital filter take the place of the analog filter, the input data must first be sampled at periodic intervals, and these samples must be input into the digital filter (computer program). The output will be in the form of discrete data values, which can be printed or used immediately in some other program or converted back to an electrical signal that is smoothed out by filtering. Thus an analog-to-digital converter is required at the input, and a digital-to-analog converter is required at the output; these are known as A/D and D/A circuits. The output of the A/D circuit could be stored in a file for use as input to the computer at a later time and processed at the speed at which this computer normally runs, or the computer could process the data immediately and just as fast as they are being generated (in real time). Present-day computers are fast enough to make this practical in most cases. Thus with A/D and D/A converters and a microprocessor in one package, it might be difficult to distinguish between a digital filter and an analog filter, since they do an equivalent job. This is called a **digital simulation**. Figure 8.1 shows a system block diagram of an analog system and its digital system replacement. We will not be discussing any of the details of the A/D and D/A circuits, but assume that somehow these circuits produce the required samples at the appropriate intervals of time, for use as digital system input, or can convert the digital system output to an accurate analog output.

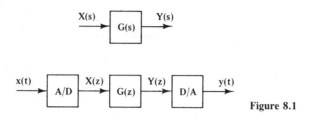

Figure 8.1

The analog active filter discussed in Chapters 5 and 6 consisted of adders, amplifiers, and integrators. In the digital filter being discussed here, the system will consist of adders, amplifiers, and delay elements. A delay element is a device that takes an input and passes it through to its output unchanged, except that it is delayed in time by one unit of time (clock period). If there are several delay elements in the circuit, the delay is the same time period for each element. The circuit symbol for the delay element is $\rightarrow \boxed{z^{-1}} \rightarrow$. The reason for choosing z^{-1} will become apparent in the next section.

Consider the circuit shown in Figure 8.2a, which is fed by input, $v_1(t)$, which is a unit step, $u(t)$, as shown in Figure 8.2b. The A/D converter starts sampling $v_1(t)$ at $t = 0$. The equal interval samples, $v[nT]$, are shown in Figure 8.2c and are tabulated in Figure 8.2d. Since $v_o[0]$ is the value of v_2 one time interval before $t = 0$, and all values were zero prior to $t = 0$, $v_o[0]$ must be zero. The rest of the values in the table can be filled in by noting that $v_o[nT] = v_2[(n - 1)T]$ and $v_2[nT] = v_1[nT] + v_o[nT]$. A plot of $v_o[nT]$ is shown in Figure 8.2e. A computer program to generate and print the values in the table is shown in Figure 8.2f. If the output, $v_o[nT]$ is put into a D/A converter, the output is smoothed and would look like Figure 8.2g. Comparing the analog input with the analog output, it is seen that this system looks like an integrator, which it actually is.

n	$v_1[nT]$	$v_o[nT]$	$v_2[nT]$
0	1	0	1
1	1	1	2
2	1	2	3
3	1	3	4
4	1	4	5
5	1	5	6
6	1	6	7
7	1	7	8
8	1	8	9

d)

```
10 V2=0: V1=1
20 V0=V2: V2=V1+V0
30 PRINT V1,V0,V2
40 GOTO 20
```

Example 8.1

Show that the circuit shown in Figure E8.1a is a differentiator by testing it with the input samples shown in Figure E8.1b.

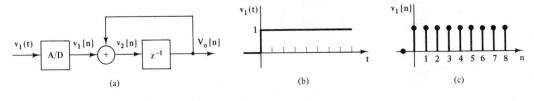

(a)   (b)   (c)

n	$v_1[nT]$	$v_0[nT]$	$v_2[nT]$
0	1	0	1
1	1	1	2
2	1	2	3
3	1	3	4
4	1	4	5
5	1	5	6
6	1	6	7
7	1	7	8
8	1	8	9

(d)

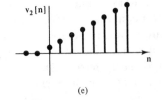

(e)

```
10 V2=0:  V1=1
20 V0=V2:  V2=V1+V0
30 PRINT V1,V0,V2
40 GOTO 20
```

(f)

(g)

Figure 8.2

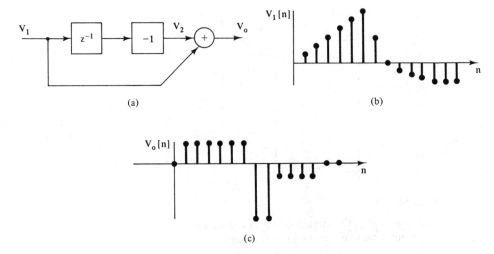

(a)   (b)

(c)

Figure E8.1

Solution Construct the table below and fill in $v_1[nT]$ by reading the plot above. Letting $v_2[0] = 0$ and seeing that $v_2[nT] = -v_1[(n - 1)T]$ and that $v_o[nT] = v_1[nT] + v_2[nT]$, complete the remainder of the table.

n	0	1	2	3	4	5	6	7	8	9	10	11	12	13	14
$v_1[nT]$	0	1	2	3	4	5	6	3	0	-0.5	-1	-1.5	-2	-2	-2
$v_2[nT]$	0	0	-1	-2	-3	-4	-5	-6	-3	0	0.5	1	1.5	2	2
$v_0[nT]$	0	1	1	1	1	1	1	-3	-3	-0.5	-0.5	-0.5	-0.5	0	0

Plotting $v_o[nT]$ in Figure E8.1c shows that this circuit really does differentiation.

8.2 z-TRANSFORM OF A FUNCTION OF TIME

The variable, s, was introduced in analog systems, so as to convert the equations to a mathematical system that was easier to solve. In discrete, or digital, systems, we introduce a new variable, z, in such a way as to make the relations between the samples and their delayed versions easier to solve. If samples of a time function, $f(t)$, are taken every T seconds, this sampled time function can be thought of as the sum of many impulse functions, each occurring at a different time and each with possibly a different height. Thus this function, $f_s(t)$, is

$$f_s(t) = f[0]\delta(t) + f[T]\delta(t - T) + \cdots + f[nT]\delta(t - nT) + \cdots \qquad (8.1)$$

The brackets indicate that these are magnitudes evaluated at discrete times. Since these are shifted impulses, we can take the Laplace transform of each term, by using the Laplace shifting theorem:

$$f_s(t) = f[0] + f[T]e^{-sT} + \cdots + f[nT]e^{-snT} + \cdots \qquad (8.2)$$

$$= \sum_{n=0}^{\infty} f[nT](e^{-snT}) = \sum_{n=0}^{\infty} f[nT]z^{-n} \qquad (8.3)$$

if

$$z = e^{sT} \qquad (8.4)$$

With this definition of z we see that the Laplace transform of this sampled function is now a function of z, not of s. So we will call this Laplace transform a z-transform instead, $F(z)$. Because only negative powers of z occur in the defining equation, Eq. 8.3, it is called the unilateral z-transform.

In addition to the unilateral character, there are other similarities between Laplace transforms and z-transforms. They are:

1. Integration and summation are very similar processes.

2. Each time function (or set of time samples) has only one transform, and vice versa.

3. Linearity applies: $a_1 f_1 + a_2 f_2 \rightarrow a_1 F_1 + a_2 F_2$.

4. Product rule will not work: the transform of a product of two functions will not equal the product of the two transforms.

The sequence, $u[nT]$, represents the samples of the step function, $u(t)$. Therefore, all the samples, $u[nT]$, are equal to one (if n is nonnegative). The delta sequence, $\delta[nT]$, is such that all samples are zero, except $\delta[0]$ is 1, not infinity. This is because the original function of time, $\delta(t)$, could have been approximated by a pulse, 1 unit high and 1 unit wide. When sampled, all the samples are zero, except at $t = 0$. Substituting this information into Eq. 8.3, we see that the z-transform of $\delta[nT]$ is 1:

$$\delta[nT] \longleftrightarrow 1 \tag{8.5}$$

It is just as easy to find the z-transform of $\delta[(n - k)T]$, which is a unit pulse (or impulse), which occurs after the kth time period. Only the kth sample is 1, so that $\delta[(n - k)T] \rightarrow z^{-k}$.

The sequence, $u[nT]$, can be considered as the sum of a series of impulses, so it is just the sum of the transforms of all those impulses (by the linearity property), or it could be found by just inserting the value, 1, into all terms of Eq. 8.3. Using the series from Section A.2, we find

$$u[nT] \rightarrow \sum_{n=0}^{\infty} z^{-n} = \frac{z}{z - 1} \tag{8.6}$$

Some examples of z-transform pairs are shown in Table 8.1. The letter, T, represents the time between samples; thus $nT = t$, the time, which is used in the original analog function that is being sampled.

The most important z-transform theorem is the shifting theorem, because a delay element is nothing more than a shifting of a given set of samples by one time period. Let $f[nT]$ be a given sequence, and $f_D[nT]$ be the delayed version. Thus

$$f_D[nT] = f[(n - 1)T] \tag{8.7}$$

The transform of the right side of Eq. 8.7 is

$$\begin{aligned} F_D(z) &= f[0 - 1] + f[1 - 1]z^{-1} + f[2 - 1]z^{-2} + f[3 - 1]z^{-3} + \cdots \\ &= f[-1] + f[0]z^{-1} + f[1]z^{-2} + f[2]z^{-3} + \cdots \end{aligned} \tag{8.8}$$

Noting that $f[-1]$ is zero, because it is before our starting time, and factoring out z^{-1}, we get

$$F_D(z) = z^{-1}(f[0]z^0 + f[1]z^{-1} + f[2]z^{-2} + \cdots) = z^{-1}F(z) \tag{8.9}$$

Thus, if $f[-1] = 0$, as is usually the case, it is seen that the transform of the delay element's output is just z^{-1} times the transform of its input. Now we see why the delay element had a z^{-1} inside its block. This delay operation, as well as seven other often-used z-transform operations, are listed in Table 8.2.

TABLE 8.1 *z*-TRANSFORM PAIRS

f[nT]	Plot of samples	F(z)	f[nT]	Plot of samples	F(z)
$\delta[nT]$		1	$\delta[(n-m)T]$		z^{-m}
$u[nT]$		$\dfrac{z}{z-1}$	$u[(n-m)T]$		$\dfrac{z}{z-1}z^{-m}$
nT		$\dfrac{zT}{(z-1)^2}$	$\dfrac{n(n-1)T^2}{2}$		$\dfrac{z}{z-1}$
a^{nT}		$\dfrac{z}{z-a^T}$	$nTa^{(n-1)T}$		$\dfrac{z}{(z-a^T)^2}$
$\cos n\omega T$		$\dfrac{z^2 - z\cos(\omega T)}{z^2 - 2z\cos(\omega T) + 1}$	$\sin n\omega T$		$\dfrac{2\sin\omega T}{z^2 - 2z\cos(\omega T) + 1}$
$(nT)^2$		$T^2\,\dfrac{z^2 + z}{(z-1)^3}$	$\dfrac{n(n-1)\cdots(n-m+1)}{m!}\,a^{(n-m)T}$		$\dfrac{z}{(z-a^T)^{m+1}}$
$e^{-anT}\cos n\omega T$		$\dfrac{z^2 - ze^{-aT}\cos}{z^2 - 2ze^{-aT}\cos\omega T + e^{-2aT}}$	$e^{-anT}\sin n\omega T$		$\dfrac{ze^{-aT}\sin\omega T}{z^2 - 2ze^{-aT}\cos\omega T + e^{-2aT}}$

TABLE 8.2 *z*-TRANSFORM OPERATIONS

Operation	Time samples	*z*-Transform
Definition	$f[nT]$	$F(z) = \displaystyle\sum_{n=0}^{\infty} f[nT]z^{-n}$
Linearity	$af_1[nT] + bf_2[nT]$	$aF_1(z) + bF_2(z)$
Delay (right shift)	$f[(n-m)T]u[(n-m)T]$	$z^{-m}F(z)$
Advance (left shift)	$f[(n+m)T]u[nT]$	$z^m F(z) - z^m \displaystyle\sum_{i=0}^{m-1} f[i]z^{-i}$
Multiply by a^{nT}	$a^{nT}f[nT]$	$F(za^{-T})$
Multiply by nT	$nTf[nT]$	$-Tz\dfrac{d}{dz}F(z)$
Initial value (first sample)		$f[0]$ equals $\lim_{z\to\infty} F(z)$
Second sample		$f[T]$ equals $\lim_{z\to\infty} [z(F(z) - f[0])]$
Final value		$f[\infty]$ equals $\lim_{z\to 1} [(1 - z^{-1})F(z)]$

Example 8.2

Find the z-transform of a square wave that has a period of 2 seconds if it is sampled every second.

Solution   The signal and its samples are shown in Figure E8.2. $F(z)$ is found directly by substituting values into Eq. 8.3, yielding

$$F(z) = 1 - z^{-1} + z^{-2} - z^{-3} + \cdots$$

This is a series, which in closed form is $F(z) = z/(z + 1)$.

An alternative method of solution is to have noticed that $f[n] = (-1)^n$, which is of the form a^{nT}, whose transform from the table is $z/(z - a^T)$ or $z/(z + 1)$.   If the original time function was a cosine wave whose period was 2 seconds, it would have identical samples, and therefore the identical z-transform.

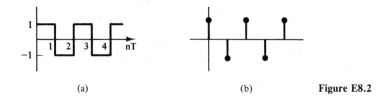

(a)   (b)   **Figure E8.2**

Example 8.3

Find the z-transform of a cosine wave that has a period of 2 seconds but is sampled every 0.5 second.

Solution   In this case the samples are as shown in Figure E8.3, and the cosine transform pair from Table 8.1 is used.   Because the wave's period is 2 seconds, ω must be π. The sampling period, T, is 0.5 second.   Putting these values into $F(z)$ yields $F(z) = z^2/(z^2 + 1)$.

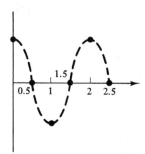

Figure E8.3

Example 8.4

Find the z-transform of the triangular wave shown in Figure E8.4a; sample period is 0.5 second.

Solution   As shown in Figure E8.4b, the samples are identical to those in Example 8.3, so the transform is the same, even though the functions being sampled are different.

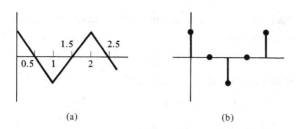

(a)   (b)   **Figure E8.4**

Example 8.5

Find $F(z)$ if $f[n] = b^{2n}$.

Solution   This is not in Table 8.1; however, if we let $a = b^2$, we have a^n, whose transform is $z/(z - a)$. Resubstituting, $F(z) = z/(z - b^2)$.

The initial value theorem in Table 8.2 is very easily proven by substituting infinity for z into Eq. 8.3.   All terms become zero, except the first.   If $F(z)$ is a ratio of two polynomials in z [such as $F(z) = N(z)/D(z)$], the degree of $N(z)$ cannot be greater than the degree of $D(z)$; if it were, $F(\infty)$ would be infinite, but the initial value theorem says that $F(\infty)$ is $f[0]$, which is certainly not infinite.   Thus if we are required to take the inverse z-transform, we will be dealing with functions of z, whose numerator degree equals its denominator degree, or functions whose numerator degree is less than its denominator degree.   Examination of Table 8.1 shows that with the exception of multiple poles and shifted functions, the numerator and the denominator of $F(z)$ are of the same degree.   If the method of partial fraction expansion of Section 1.9 were applied, it would yield fractions with a numerator one degree less than the denominator usually, a constant numerator and a binomial denominator.   This would be an unsatisfactory condition as far as use of Table 8.1 is concerned.   To get around this problem, we first divide both sides by z, then expand the right side by partial fractions, giving a one-degree difference between numerator and denominator.   Then multiplying through by z will make the degree of the numerator of $F(z)$ the same as that of the denominator.

Example 8.6

Find $f[n]$ if $F(z) = (2z^3 - 5z^2 + 7)/z^3$.

Solution   Doing the indicated division on the right side, we get

$$F(z) = 2 - \frac{5}{z} + \frac{7}{z^3} \longrightarrow 2\delta[n] - 5\delta[n - 1] + 7\delta[n - 3]$$

Example 8.7

Find the formula for $f[nT]$ and the first six samples if

$$F(z) = \frac{z - 1}{(z - 2)(z - 3)^2}$$

Solution Dividing through by z and expanding into partial fractions, we get

$$\frac{z - 1}{z(z - 2)(z - 3)^2} = \frac{A}{z} + \frac{B}{z - 2} + \frac{C}{z - 3} + \frac{D}{(z - 3)^2}$$

Using the methods of Sections 1.9 and 1.10, we find

$$A = \tfrac{1}{18} \quad B = \tfrac{1}{2} \quad C = -\tfrac{5}{9} \quad D = \tfrac{2}{3}$$

Now multiplying through by z to restore $F(z)$, we get

$$F(z) = \frac{1}{18} + \frac{(\tfrac{1}{2})z}{z - 2} + \frac{(-\tfrac{5}{9})z}{z - 3} + \frac{(\tfrac{2}{3})z}{(z - 3)^2}$$

The inverse z-transforms can now be taken from Table 8.1.

$$f[nT] = (\tfrac{1}{18})\delta[nT] + (\tfrac{1}{2})2^n - (\tfrac{5}{9})3^n + (\tfrac{2}{3})nT(3)^{n-1}$$

Substituting 0, 1, 2, 3, 4, 5 for n, and assuming that $T = 1$, we obtain the first six samples: $f[n] = 0, 0, 1, 7, 35, 151$.

Example 8.8

Use the shifting operation to find $f[nT]$ for

$$F(z) = \frac{2 + 5z^{-1} + 3z^{-3}}{z^2 - 2z + 1}$$

Solution Let $W(z) = z/(z^2 - 2z + 1) = z/(z - 1)^2$; then $w[nT] = nT$ and

$$F(z) = W(z)(2z^{-1} + 5z^{-2} + 3z^{-4}) = 2z^{-1}W(z) + 5z^{-2}W(z) + 3z^{-4}W(z)$$

Using the shift operation on each term yields

$$f[nT] = 2(n - 1)Tu(nT - T) + 5(n - 2)Tu(nT - 2T) + 3(n - 4)Tu(nT - 4T)$$

8.3 USING THE z-TRANSFORM TO ANALYZE DIGITAL FILTERS

In analysis we presume that we have a digital circuit or at least a computer program that is in essence a digital circuit. Our job then is to analyze this circuit to see how it behaves; that is, if an input sequence is specified, what is the output sequence? For instance, if a computer program is specified, what will be its output? For our purposes it is best if the computer program is first put into flowchart form. We will be concerned only with mathematical statements; that is, we will ignore strings and graphics and read and write statements and rounding routines. Another restriction will be that all the operations must be linear; that is, no exponentiation or sine functions or absolute value operations will be allowed within the program, except in defining the input samples. Our procedure will be to convert this flowchart into a z-transform block diagram, if necessary. From this a z-transform equation is written (in most cases this equation is gotten directly from the flow-

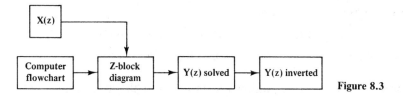

Figure 8.3

chart).   This is then solved after inserting the *z*-transform of the input; finally, $Y(z)$ is inverted to find the output sequence.   This is illustrated in Figure 8.3.

Example 8.9

Find the output samples formula for the computer program which is flowcharted as shown in Figure E8.9 if the input is $n2^{n-1}$.

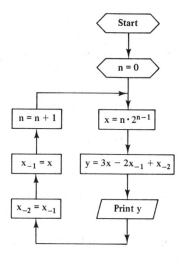

Figure E8.9

Solution   We can write the *z*-transform equation directly from the flowchart:

$$Y(z) = 3X(z) - 2z^{-1}X(z) + z^{-2}X(z)$$

$$= X(z)\,\frac{3z^2 - 2z + 1}{z^2}$$

From Table 8.1, $n(2^{n-1}) \rightarrow z/(z-2)^2$; substituting and dividing by z gives

$$\frac{Y(z)}{z} = \frac{3z^2 - 2z + 1}{z^2(z-2)^2} = \frac{A}{z^2} + \frac{B}{z} + \frac{C}{(z-2)^2} + \frac{D}{z-2}$$

where $A = \frac{1}{4}$, $B = -\frac{1}{4}$, $C = \frac{9}{4}$, and $D = \frac{1}{4}$, using Eq. 1.21.   Therefore,

$$Y(z) = \frac{\frac{1}{4}}{z} - \frac{1}{4} + \frac{9z/4}{(z-2)^2} + \frac{z/4}{z-2}$$

The inverse of this is (using Table 8.1)

$$y[nT] = (\tfrac{1}{4})\, \delta[n-1] - (\tfrac{1}{4})\, \delta[nT] + (\tfrac{2}{4})n2^{n-1} + (\tfrac{1}{4})2^{n}$$

The first six values of $y[nT]$ are (if $T = 1$) 0, 3, 10, 29, 76, 188.

In Example 8.9 the output equation was in terms of the input values only. This is called a nonrecursive equation. The following example illustrates a recursive equation (i.e., an equation whose output depends not only on the samples of the input), but also on previous values of the output (i.e., feedback).

Example 8.10

Analyze the circuit in Figure E8.10, finding the time expression corresponding to the output sequence, if the input is samples of $u(t)$.

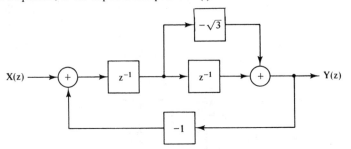

Figure E8.10

Solution   Looking at the rightmost summing junction, we see that

$$Y(z) = (-Y(z) + X(z))z^{-2} + (-Y(z) + X(z))z^{-1}(-\sqrt{3})$$

$$Y(z)(1 - \sqrt{3}\, z^{-1} + z^{-2}) = X(z)(-\sqrt{3}\, z^{-1} + z^{-2})$$

Solving for $Y(z)/z$ and putting $z/(z-1)$ for $X(z)$ (Table 8.1) yields

$$\frac{Y(z)}{z} = \frac{-\sqrt{3}\, z + 1}{(z^2 - \sqrt{3}\, z + 1)(z-1)} = \frac{-1 - \sqrt{3}}{z - 1} + \frac{(1 + \sqrt{3})z - (2 + \sqrt{3})}{z^2 - \sqrt{3}\, z + 1}$$

The above is by partial fraction expansion. Now multiply by z:

$$Y(z) = (-1 - \sqrt{3})\frac{z}{z-1} + (1 + \sqrt{3})\frac{z^2 - z(\sqrt{3}/2) - z(1/2)}{z^2 - 2z(\sqrt{3}/2) + 1}$$

Referring to Table 8.1, we see that the first term is the constant, -2.732, and the second term is a combination of sine and cosine terms with $\omega T = \pi/6$ radians (since $\cos \omega T = \sqrt{3}/2$ and $\sin \omega T = \tfrac{1}{2}$). Thus the entire solution is

$$y[nT] = -2.732[u[nT] - \cos(n\omega T) + \sin(n\omega T)]$$

$$= -2.732[u[nT] + \sqrt{2} \sin(n\omega T - 45°)]$$

If it is desired to have this circuit oscillate at $f = 1$ kHz, then $\omega = 6280$ rad/s and the period is 1 ms. But since $T = \pi/6\omega$, the sampling period must be 0.083 ms, or 12 samples per cycle. Thus, for $t > 0$,

$$y(t) = -2.732[1 + \sqrt{2} \sin(6280t - 45°)]$$

Example 8.10 illustrated that recursive filters involve a feedback of signals and the possibility of oscillation. Notice that although the input was dc of 1 V, the circuit oscillated with a peak-to-peak value of 7.734 V. This means that the system is marginally stable. By observing the transfer function, $G(z) = Y(z)/X(z)$, you will be able to determine the stability, just as you did for analog circuits; however, the rules are somewhat different. After finding $G(z)$, factor it and cancel common factors between numerator and denominator, if there are any. Then make a list of all the poles. The system is unstable if there are any poles whose magnitude is greater than 1 or if there are multiple poles whose magnitude equals 1. If there are one or more simple poles whose magnitude is 1, the system is marginally stable. If none of the above apply, the system is stable. In example 8.10,

$$G(z) = \frac{Y(z)}{X(z)} = \frac{-\sqrt{3}\,z + 1}{z^2 - \sqrt{3}\,z + 1} = \frac{-1.732z + 1}{(z - 0.866 + j0.5)(z - 0.866 - j0.5)}$$

Therefore, poles are $0.866 + j0.5$ and $0.866 - j0.5$. Converting these to polar form, it is seen that the magnitude of both are 1, although they are different poles. So it is marginally stable, as already mentioned.

Many times it is desired to find the frequency response of a circuit, that is, the steady-state response of a digital circuit to samples of a sinusoidal input at any arbitrary frequency. The gain magnitude formula is

$$A(\omega) = \left|[G(z)]_{z=e^{j\omega T}}\right| \tag{8.10}$$

Algebraic manipulation using Euler's formulas A1.5 and A1.6 or A1.8 of Appendix A must be performed just prior to taking the absolute magnitude indicated in Eq. 8.10.

Example 8.11

Find the frequency response, $A(\omega)$, of the digital filter in Example 8.10.

Solution   Using Eq. 8.10 on the $G(z)$ of Example 8.10, we have

$$\frac{-\sqrt{3}\,e^{j\omega T} + 1}{e^{j\omega 2T} - \sqrt{3}\,e^{j\omega T} + 1}$$

Since the first and last terms of the denominator have like coefficients (1), it is convenient to employ formula A1.5 of Appendix A. Since this is not the case with the numerator, it is best to use formula A1.8. The algebra is as follows:

$$\left|\frac{-\sqrt{3}e^{j\omega T} + 1}{e^{j\omega T}(e^{j\omega T} - \sqrt{3} + e^{-j\omega T})}\right| = \left|\frac{-\sqrt{3}(\cos\omega T + j\sin\omega T) + 1}{e^{j\omega T}(2\cos\omega T - \sqrt{3})}\right|$$

$$= \frac{\sqrt{(-\sqrt{3}\cos\omega T + 1)^2 + (\sqrt{3}\sin\omega T)^2}}{2\cos\omega T - \sqrt{3}} = \frac{\sqrt{4 - 2\sqrt{3}\cos\omega T}}{2\cos\omega T - \sqrt{3}}$$

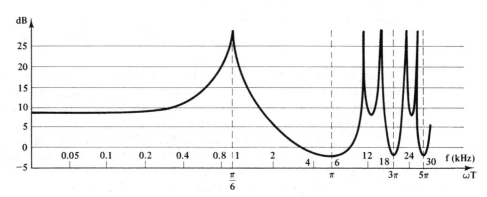

Figure E8.11

Substituting $T = 0.0000833$ gives us

$$A(\omega) = \frac{\sqrt{4 - 3.464 \cos (0.0000833 \, \omega)}}{2 \cos (0.0000833 \, \omega) - 1.732}$$

Entering this equation into a programmable calculator, inputting several values of ω, converting the result to decibels and plotting on log frequency paper, the plot shown in Figure E8.11 is obtained.   Note that at $\omega = 6283$ (i.e., 1 kHz), the filter resonates. This corresponds to the pole of the $G(z)$ function.   Note also that at $\omega T = \pi$, 3π, 5π, and so on, the response repeats; that is, the frequency response itself is periodic. This is something that never happened in analog systems but always happens in digital systems.

Comparing recursive and nonrecursive filters, it is found that a larger number of elements is required to achieve a sharp cutoff in a nonrecursive filter, although it has excellent phase characteristics.   Because the nonrecursive filter's output depends on the present input sample and a finite number of the input's previous samples (because the circuit contains a finite number of delay elements), the impulse response will fall to zero after m delay periods if there are m delay elements; thus this type of filter is sometimes called a finite-impulse-response (FIR) filter.   All nonrecursive filters are FIR; but it is possible to build an FIR filter using a recursive filter.   The opposite of FIR is IIR, which means infinite impulse response, which can be achieved only using a recursive filter.   This means that after the input is shut off, an output will be present forever (decaying, stable, or growing with time).

8.4 DIGITAL SIMULATION OF THE ANALOG TRANSFER FUNCTION

There are several methods of synthesizing digital filters.   In the Example 8.11, $A(\omega)$ was a function of the cosine function.   This occurs quite often in digital filters, because $z = e^{j\omega T}$ (for steady state) and e leads to sinusoids (see formulas A1.5 and A1.6 of Appendix A).   So Example 8.11 is an IIR filter, but an FIR filter simulation (no feedback) can also be built and also involves sinusoids. These "sinusoidal" responses have no counterpart to the analog case; that is, they are not Butterworth, Chebyshev, and so on.

We will not consider this type in this section. Instead, we will try to make digital simulations of the analog functions that have been studied previously, such as Butterworth, and so on. The starting point for these methods is to get a Butterworth or Chebyshev transfer function, $G(s)$. The end product will be a IIR filter, instead of FIR. One method, call it the frequency transformation method, requires that s be replaced by a digital subcircuit that performs differentiation, since s means differentiation. A digital circuit that does this will have a z-transfer function of

$$G(z) = \frac{2}{T} \frac{1 - z^{-1}}{1 + z^{-1}} \tag{8.11}$$

This is only an approximate differentiator and the error becomes quite large at higher frequencies. By choosing T small enough, this error can be reduced in the range of frequencies of interest. The procedure for choosing T is as follows: Decide on the highest frequency (called ω_{acc}, rad/s) at which you require accuracy. Depending on the accuracy you require, use one of the following formulas for T:

$$T = \frac{0.31}{\omega_{acc}} \,(1\%) \qquad T = \frac{0.45}{\omega_{acc}} \,(2\%) \qquad T = \frac{1}{\omega_{acc}} \,(10\%) \tag{8.12}$$

It is not necessary to build this subcircuit for each time that s occurs in the $G(s)$, because simplification might occur if Eq. 8.11 is just substituted into the $G(s)$ and simplified algebraically. Then the entire $G(z)$ can be realized by cross-multiplying and constructing the block diagram. The following example will illustrate this technique.

Example 8.12

Using the frequency transformation technique, construct a block diagram for a digital filter that will simulate a first-order Butterworth filter that has a 3-dB cutoff at 100 Hz and is accurate within 10% up to 200 Hz.

Solution   Since 100 Hz is 628 rad/s, the analog transfer function is $1/(s + 628)$. Since 10% accuracy is desired up to 1256 rad/s, T will be 0.000796 second. Substituting Eq. 8.11 for s in the $G(s)$ yields

$$G(z) = \frac{1}{\dfrac{2}{0.000796} \dfrac{1 - z^{-1}}{1 + z^{-1}} + 628} = \frac{1 + z^{-1}}{3140 - 1884z^{-1}} = \frac{Y(z)}{X(z)}$$

Cross-multiplying, we obtain

$$3140Y - 1884z^{-1}Y = X + z^{-1}X$$

or

$$Y = 0.0003185X + z^{-1}(0.6Y + 0.0003185X)$$

The block diagram form of this is shown in Figure E8.12.

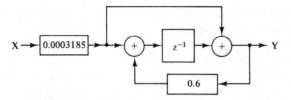

Figure E8.12

Another method for simulating an analog transfer function is to use partial fraction expansion on the analog transfer function, $G(s)$, and then find a $G(z)$ for each partial fraction. These can then be recombined into one fraction in z. Then by cross-multiplying, we can construct the block diagram or computer program. Let us call this the time-sampling method.

The critical part is to find the proper $G(z)$ for each partial fraction. The idea in any digital simulation is to have the samples of digital output identical to the output values of the corresponding analog filter if that analog filter had only samples of the normal analog input as its input.

If the analog transfer function had only simple real poles, then, using partial fraction expansion,

$$G(s) = \sum \frac{A_i}{s - p_i} \tag{8.13}$$

It can be shown that the z-transform will be

$$G(z) = \sum \frac{A_i T}{1 - e^{p_i T} z^{-1}} \tag{8.14}$$

This will be a good approximation if the largest p value times T yields a value of 0.1 or less. Each one of these fractions can be simulated by a digital system and the outputs of all these systems can be summed at a final adder to give the total output; but a better way to handle this is to recombine all the partial fractions of Eq. 8.14 into one big fraction again and then simulate this fraction with one digital system. Notice that there is no way of making a substitution in the original $G(s)$ to get the $G(z)$; we must do the partial fraction expansion and recombination outlined above. Also, if the $G(s)$ is stable, the $G(z)$ will also be stable. The following example will illustrate this technique.

Example 8.13

Do a digital simulation of

$$G(s) = \frac{2s}{(s + 2)(s + 4)}$$

Using the time-sampling method if it is to be accurate within 2% up to 2 Hz.

Solution According to Eq. 8.12, the sampling period, T, is 0.0716 second. However, Eq. 8.12 applies only to the frequency transformation method. For the time-sampling method, the largest pole is 4. So $4T$ must be much smaller than 1. Let us choose $T = 0.02$ second. By partial fraction expansion,

$$G(s) = \frac{-2}{s + 2} + \frac{4}{s + 4}$$

Therefore, using Eq. 8.14,

$$G(z) = \frac{-2(0.02)}{1 - e^{-2(0.02)}z^{-1}} + \frac{4(0.02)}{1 - e^{-4(0.02)}z^{-1}}$$

$$\frac{Y(z)}{X(z)} = G(z) = \frac{-0.04}{1 - 0.961z^{-1}} + \frac{0.08}{1 - 0.923z^{-1}} = \frac{0.04 - 0.04z^{-1}}{1 - 1.884z^{-1} + 0.887z^{-2}}$$

Cross-multiplying and putting all terms but Y on the right side gives us

$$Y = 0.04X + z^{-1}[1.884Y - 0.04X - z^{-1}(0.887Y)]$$

This yields the block diagram and computer program shown in Figure E8.13.

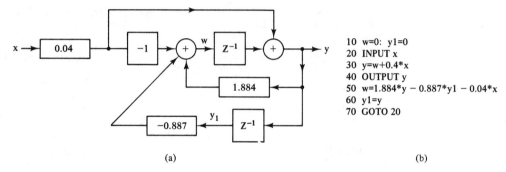

```
10  w=0:  y1=0
20  INPUT x
30  y=w+0.4*x
40  OUTPUT y
50  w=1.884*y − 0.887*y1 − 0.04*x
60  y1=y
70  GOTO 20
```

(a)   (b)

Figure E8.13

Thus a disadvantage of this method, at least in Example 8.13, is that more samples must be taken each second to get the same accuracy as for the frequency transformation method. In the next example we make a direct comparison of these two methods and compare them with the analog circuit response.

Example 8.14

Repeat Example 8.12 with the same sampling period, $T = 0.000796$ second, but use the time-sampling method, and compare the methods.

Solution   Since $G(s) = 1/(s + 628)$, $G(z) = 0.000796/(1 - e^{-628(0.000796)}z^{-1})$, using Eq. 8.14. So

$$G(z) = \frac{Y(z)}{X(z)} = \frac{0.000796}{1 - 0.6066z^{-1}} \quad \text{or} \quad Y = 0.000796X + 0.6066Yz^{-1}$$

The block diagram is shown in Figure E8.14. The gain function

$$A(\omega) = \left| \frac{0.000796}{1 - 0.6066e^{-j\omega T}} \right|$$

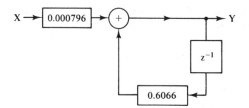

Figure E8.14

Using formula A1.8, this is

$$\frac{0.000796}{1 - 0.6066 \cos \omega T + j0.6066 \sin \omega T}$$

Converting the denominator to polar form and dropping the angle yields

$$A(\omega) = \frac{0.000796}{\sqrt{1.368 - 1.2132 \cos (0.000796\omega)}} \quad \text{time-sample method}$$

Substituting $e^{j\omega T}$ for z in the $G(z)$ of Example 8.12 and applying formula A1.8 yields the gain function for that method:

$$A(\omega) = \frac{2 \cos (0.000398\omega)}{10^3\sqrt{13.410 - 11.830 \cos (0.000796\omega)}} \quad \text{frequency transformation method}$$

The analog gain is

$$A(\omega) = \frac{1}{\sqrt{\omega^2 + 628^2}} \quad \text{analog method}$$

The following table shows the gains for all three methods in decibels at selected frequencies.

f	A.M.	F.T.M.	T.S.M.
0	−56	−56	−53.9
10	−56	−56	−53.9
100	−58.9	−59.1	−56.8
200	−63	−63.6	−60.5
300	−66	−67.7	−63.1
628	−72	−137	−66.1
956	−75.7	−67.7	−63.1

Note the agreement between A.M. and F.T.M. Also, note that the digital methods fold about 628 Hz frequency.

Although the time-sampling method in many problems produces a greater error than the frequency transformation method, as we saw from Example 8.14, we also see that the error is fairly constant over the band of interest. In many cases this is acceptable, or we can just multiply the output by a suitable constant to adjust for this phenomenon.

In the case of band-reject filters or high-pass filters, a problem arises for the time-sampling method. Since the $G(s)$ for these filters has the same degree in the denominator as in the numerator, it is impossible to use partial fraction expansion to get just the terms indicated by Eq. 8.14. There will always be an additional constant term. This can be avoided by using the step response instead of the impulse response. The Laplace transform of the step response is $G(s)/s$. This has the effect of reducing the degree of the numerator, so that Eq. 8.14 can be used, but without using the T in the numerator. The function of z that results from this is then the z-transform of the digital step response. Dividing this by the transform of the step input, $z/(z - 1)$, will then give the $G(z)$ for what is called the step-invariant approximation of the analog function. The following example illustrates its use.

Example 8.15

Use the time-sampling method to find $G(z)$, to match $G(s) = 2s/(s + 1000)$, up to about 1 kHz.

Solution Since Eq. 8.14 will not work in this case, convert the analog step output to the digital step output, rather than the analog transfer function to the digital transfer function.

$$\frac{G(s)}{s} = \frac{2}{s + 1000}$$

so

$$G(z)\frac{z}{z - 1} = \frac{2}{1 - e^{-1000T}z^{-1}}$$

Notice that no T appears in the numerator, since this is a step-invariant transformation time-sampling method. Since the pole is 1000 and pT should be about 0.1, pick $T = 0.0001$. The folding frequency will then be $\pi/0.0001 = 31{,}416$ rad/s. Substituting $T = 0.0001$ in the equation above and solving yields

$$G(z) = \frac{2z - 2}{z - 0.905} \qquad A(\omega) = \left| \frac{2z - 2}{z - 0.905} \right|_{z = e^{j\omega 0.0001}}$$

Using the trig conversions,

$$A(\omega) = \frac{4 \sin (0.00005\omega)}{\sqrt{1.819 - 1.81 \cos (0.0001\omega)}}$$

The following table compares decibel values at various frequencies for $A(\omega)$ for this digital filter and for the analog $(= 2\omega/\sqrt{\omega^2 + 1000^2})$.

f (Hz)	Digital	Analog
1	-37.6	-38.0
10	-17.6	-18.0
159	$+3.4$	$+3.0$
318	$+5.5$	$+5.1$
1000	$+6.3$	$+5.9$
5000	$+6.4$	$+6.0$

Thus we see that the digital filter is 0.4 dB higher than the analog filter all the way to the "folding" frequency, 5 kHz. At 9 kHz it would drop to 6.3 dB; at 9990 Hz, it drops to -17.6 dB.

An all-pass digital filter can also be synthesized by the procedure above, although the phase response would be quite poor. Since the reason for using an all-pass filter is to get a linear phase response, this would seem to rule out digital filters for this purpose. However, FIR filters have very good phase response; and of course, the simplest way to realize an all-pass filter is with just one or more delay elements between the input and output.

Example 8.16

Design an all-pass filter to have a time delay of 3 ms for signals in the range 0 to 900 Hz.

Solution   Since the range is to 900 Hz, T must be less than or equal to $\pi/(2\pi900) = 0.555$ ms. Then 3-ms delay divided by 0.555 ms delay per section gives 5.4 sections, so 6 sections are required. To get 3 ms total, each section must have 0.50 ms delay, not 0.555. This just means the signal range is somewhat larger (1 kHz) than required. This simple digital circuit is as shown in Figure E8.16.

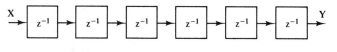

Figure E8.16

Another method of altering the frequency response is to substitute $-z$ for z in the $G(z)$. The effect is to shift the frequency response by an amount equal to the fold-over frequency. This can be considered a leftward shift or a rightward shift, since the frequency response is periodic at twice the foldover frequency, including negative frequencies for this purpose. This means that a low-pass filter designed either by the frequency transformation method or the time-sampling method can have its frequency response shifted so as to create a high-pass filter, as long as the frequencies of interest are below the foldover frequency. The disadvantages are that the results do not match those of the corresponding analog filter. For instance, using this technique, a low-pass filter shifted to make a high-pass filter will have a much steeper roll-off rate for the same number of elements, but the usable passband will be rather narrow.

The actual digital filter is realized as a computer program, that is, as software. It is also possible to realize it as hardware. One possibility is to use the MF-10 switched capacitor integrated-circuit chip, described in Chapter 7. This is so because any charge applied to the switched capacitor during one phase of the clock does not transfer (is delayed) until the next clock phase. Theory and details are not covered in this book.

SUMMARY

In this chapter a whole new system of signals was studied: digital signals. These are closely associated with a sampled analog system, that is, signals that consist of impulses equally spaced in time. Since the Laplace transform of such a signal will contain exponentials involving s for each such impulse, a substitution is conveniently made which replaces the s variable with a new variable, z. The mathematical symbol, z^{-1}, is represented physically by a delay element.

The study of digital filters involves the z-transform on the mathematical side, the block diagram on the pictorial side (instead of a schematic), and either a computer program or a dedicated microprocessor or switched capacitor IC circuit with A/D and D/A converters on the physical side.

The digital filter is limited to frequencies below one-half the sampling frequency. Thus any frequency can be handled if the sampling frequency is raised high enough. The problem is that the computer or the converters may not be fast enough to handle all the calculations in real time if information is supplied at such a high rate. Before the A/D converter, the analog signal may have to go through a low-pass filter to limit its frequency range just in case there might be some frequencies in the input that are too high.

Another problem is that if a digital filter is to simulate an analog filter function, the results are only approximate, but if the procedures are adhered to, the degree of approximation can be as good as desired and any analog filter can be duplicated in digital form.

PROBLEMS

Section 8.1

8.1. Find the z-transform transfer function $G(z)$ for Figure P8.1.

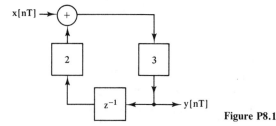

Figure P8.1

8.2. Find the z-transform transfer function $G(z)$ for Figure P8.2.

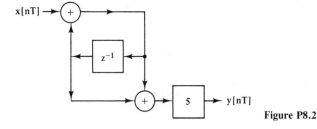

Figure P8.2

8.3. Consider the block diagram shown in Figure P8.3.
   (a) Find its transfer function, $G(z)$.
   (b) If seven such sections are cascaded instead of two, what is the transfer function, $G(z)$?

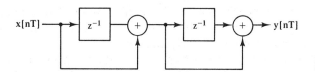

Figure P8.3

Section 8.2

8.4. Using delay elements and adders, synthesize (make a block diagram for) the transfer function

$$G(z) = \frac{z^2 - 5z + 2}{z^2}$$

8.5. Find $X(z)$ if $x[0] = 5$, $x[1] = 1$, $x[2] = 0$, $x[3] = -2$, and all others $= 0$.

8.6. Find $X(z)$ if the x samples are 0, 1, 0, 1, 0, 1, 0, and so on. (*Hint:* See Section A.2.)

8.7. Find $X(z)$ if the samples are as shown in Figure P8.7 and all other samples are zero.

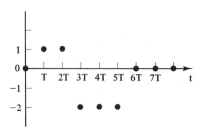

Figure P8.7

8.8. Find $X(z)$ if $x[nT] = u[nT - 2T]$.

8.9. Find $X(z)$ if $x[nT] = u[3nT] - u[nT - 2T]$.

8.10. Find $X(z)$ if $X[nT] = \cos{(nT)}$ and $T = 1$ second.

8.11. A digital circuit has the transfer function

$$G(z) = \frac{z^2 - z}{(z - 0.5)(3z - 2)}$$

The input is $u[nT - T]$.
   (a) What is the output in the z-domain?
   (b) What is the output in the t-domain?
   (c) What are the first four values of the output?
   (d) What is the final value (at $n = $ infinity)?

Section 8.3

8.12. The impulse response of a certain digital system is

$$g[nT] = 3e^{-2nT}u[nT]$$

$T = 1$ second.
(a) Find the first six samples of the impulse response.
(b) Find the formula for the step response in the z-domain.
(c) Find the formula for the step response in the t-domain.
(d) Find the first six samples of the step response.
(e) Apply the final value theorem to the answer from part (b).
(f) How does part (e) compare with part (d)?

8.13. The following $G(z)$ formulas pertain to three different systems.   Which are stable and which are unstable?

(a) $\dfrac{z - 2}{z^2 - 1.5z + 1.125}$ (b) $\dfrac{z - 2}{z^2 + 1.5z + 1.125}$ (c) $\dfrac{z - 2}{z^2 - 2.75z + 1.5}$

Section 8.4

8.14. A digital circuit has the impulse response, $y[nT] = \cos nT$. $T = 0.5$ second.
(a) Find $Y(z)$.
(b) Make a block diagram of the circuit.
(c) Write a computer program in BASIC to simulate it.
(d) What are the first five values of the output of this program?
(e) Compare these with $\cos (n/2)$ for $n = 0, 1, 2, 3, 4$.
(f) What might be a good name for this circuit?

8.15. A certain sampled function is $x[nT] = e^{-nT}u[nT]$.
(a) Find the Laplace transform of these samples.
(b) $X(z)$ can be found by substituting $[\ln(z)]/T$ for s in the Laplace transform of the impulse sequence representing samples.   Since we have found this Laplace transform in part (a), make this substitution to find $X(z)$.

8.16. An all-pass analog circuit with transfer function $G(s) = (s - 5)/(s + 5)$ is to be converted into a digital circuit.
(a) If $T = 0.01$ second, find $G(z)$ using the time-sampling method.
(b) Draw the block diagram.

8.17. Repeat Problem 8.16 using the frequency transformation method.

appendix A: useful formulas

A.1 TRIGONOMETRIC FORMULAS

A1.1. $\sin (A + B) = \sin A \cos B + \cos A \sin B$

A1.2. $\cos (A + B) = \cos A \cos B - \sin A \sin B$

A1.3. $\sin (A - B) = \sin A \cos B - \cos A \sin B$

A1.4. $\cos (A - B) = \cos A \cos B + \sin A \sin B$

A1.5. $\sin \omega t = \dfrac{e^{j\omega t} - e^{-j\omega t}}{2j}$

A1.6. $\cos \omega t = \dfrac{e^{j\omega t} + e^{-j\omega t}}{2}$

A1.7. $D \cos B + E \sin B = C \sin (A + B)$, where $A = \tan^{-1} (D/E)$ and $C = \sqrt{D^2 + E^2}$

A1.8. $e^{j\omega t} = \cos \omega t + j\sin\omega t$

A.2 SERIES

$$\frac{1}{1 + x} = 1 - x + x^2 - x^3 + x^4 - \cdots \qquad (\text{if } |x| < 1)$$

A.3 L'HÔPITAL'S RULE

To find $f(x)/g(x)$, evaluated at $x = a$, when this substitution yields an indeterminate answer, such as 0/0, or ∞/∞: Substitute $x = a$ into $f'(x)/g'(x)$. If this still does not work, try $x = a$ into $f''(x)/g''(x)$, and so on.

appendix B: computer programs

B.1 NETWORK

```
900   TEXT
950 DZ =  - 1000
1000  REM      NAME: N  E  T  W  O  R
      K
1020  REM
1040  REM      UPDATED   - 30 MAY
      84
1100  TEXT : HOME : INVERSE
1120  PRINT  TAB( 5);"N E T W O R
      K   A N A L Y S I S   "
1140  NORMAL : PRINT
1160 I = 0:J = 0:K = 0:N = 0:I1 =
      0
1180  GOSUB 6200
1190  FOR I = 1 TO 1000:II = II +
      1: NEXT I
1200 Y = 10: INPUT "HOW MANY NODE
      S (DON'T INCL. REF (#0)) ?";
      Z$
1220  IF Z$ <  > "" THEN Y =  VAL
      (Z$)
1240  DIM A(Y,Y),B(Y,Y),P(Y,Y),Q(
      Y,Y),B1(Y,Y),Q1(Y,Y),AB(200)
      ,DN(200),PN(200)
```

```
1260  PRINT "MEMORY ALLOCATED FOR
      ";Y;" NODES."
1280  D$ =  CHR$ (4): REM    CONTRO
      L D
1300  N = 0:PI = 3.141592654:LGTEN
      = 8.685889640
1320  ED = 1: REM    ALLOW EDITOR S
      UPPORT
1540  INPUT Z$
1560  E$ = Z$:Z$ =  LEFT$ (Z$,1)
1580  IF Z$ = "R" THEN 1860: REM
              RESISTORS
1600  IF Z$ = "C" THEN 1900: REM
              CAPACITORS
1620  IF Z$ = "L" THEN 1960: REM
              INDUCTORS
1640  IF Z$ = "A" THEN 2140: REM
              AMPLIFIERS
1660  IF Z$ = "F" THEN 2020: REM
              F.E.T.S
1680  IF Z$ = "N" THEN 2060: REM
              NPN TRANSISTOR
1700  IF Z$ = "E" THEN 2220: REM
              END OF INPUT
1740  INVERSE : PRINT
1760  PRINT "BAD COMPONENT TYPE:"
      ;: NORMAL : PRINT " ";: FLASH
      : PRINT Z$
1780  NORMAL
1800  PRINT
1810  PRINT "RE-DO THE LAST COMPO
      NENT"
1820  GOTO 1540
1840  END
1860  INPUT I,J,V: REM        RESIS
      TOR
1880  V = 1 / V: GOSUB 3900: GOTO
      1540
1900  INPUT I,J,V: REM        CAPAC
      ITOR
1920  V = V / 1000000
1940  GOSUB 4120: GOTO 1540
1960  INPUT I,J,V: REM        INDUC
      TOR
1980  V =  - 1 / V
2000  GOSUB 4260: GOTO 1540
2020  INPUT K,J,I,V: REM        F.E
      .T.
2040  L = J: GOTO 2200
2060  INPUT K,J,I,B1,V: REM
      NPN TRANSISTOR
2080  L = I:I = K:V = 1 / V: GOSUB
      3900
2100  I = L:L = J: GOTO 2180
2120  REM      OP AMP
2140  INPUT K,L,J,I,B1,V: REM    I
      N+,-; OUT+,-; GAIN, OHMS
2160  V = 1 / V: GOSUB 3900
2180  V = B1 * V
2200  GOSUB 4400: GOTO 1540
```

```
2220  INPUT E,F: REM     END,READ
      I/O
2260  I =  LEN (E$) - 1: IF I = 0 THEN
      E$ = " ": GOTO 2300
2280  E$ =  RIGHT$ (E$,1): REM    S
      AVE END MESSAGE
2300  FOR I = 0 TO N
2320  FOR J = 0 TO N
2340  P(I,J) = A(I,J)
2360  Q1(I,J) = B1(I,J)
2380  Q(I,J) = B(I,J)
2400  NEXT J: NEXT I
3000  PRINT
3020  PRINT "THIS CIRCUIT HAS ";N
      ;" NODES. "
3040  PRINT "NODE ";E;" IS INPUT,
      ";F;" IS OUTPUT"
3060  PRINT
3080  PRINT "DO YOU WISH TO CHANG
      E INPUT/OUTPUT NODES?";
3100  GET Z$
3120  IF Z$ <  > "Y" THEN 3200
3140  PRINT : PRINT
3160  PRINT "INPUT NODE, OUTPUT N
      ODE:";
3180  INPUT E,F
3200  REM  ENTRY POINT FOR NEW FR
      EQ RANGE
3220  HOME
3240  PRINT E$: REM    END MESSAGE

3260  PRINT
3280  PRINT "FREQUENCY RANGE FOR:
      ";F$
3300  PRINT "   START, END, INC
      (- FOR LOG)"
3320  INPUT G,H,D
3330  SD = 10 ^ ( INT (LGTEN *  LOG
      (G) / 20) - 3)
3340  PRINT
3360  PRINT "FREQ        DB
      PHASE"
3380  IF D < 0 THEN F2 =  - D: GOTO
      3420
3400  F2 =  INT (1.5 + (H - G) / D
      )
3420  IF D < 0 THEN D =  - ((H /
      G) ^ (1 / ( - D - 1)))
3440  F1 = G
3459  DL = 1000:DH =  - DL
3460  FOR I1 = 1 TO F2
3480  W = 2 * PI * F1:D1 = E:D2 =
      F: GOSUB 5660
3500  V = B1:U = D2
3520  IF ( - 1) ^ (E + F) > 0 THEN
      GOTO 3560
3540  U = U - 180
3560  D1 = E:D2 = E
3580  GOSUB 5660:V = V / B1:U = U
      - D2
```

```
3600   IF U > 180 THEN U = U - 360

3620   IF U <  - 180 THEN U = U +
       360
3638   IF V < 1E - 35 THEN V = 1E -
       35
3640 DB = LGTEN *  LOG (V)
3660   PRINT  INT (F1 / SD + .5) *
       SD;: HTAB 11
3680   PRINT ( INT (DB * 100 + .5)
       / 100);: HTAB 22
3700   PRINT ( INT (U * 10 + .5) /
       10)
3702 DN(I1) = DB:AB(I1) = F1:PN(I
       1) = U
3704   IF DN(I1) > DH THEN DH = DN
       (I1)
3706   IF DN(I1) < DL THEN DL = DN
       (I1)
3720   IF D < 0 THEN F1 =  - F1 *
       D: GOTO 3760
3740 F1 = F1 + D
3760   NEXT I1
3762 DR = DH - DL
3780   PRINT  CHR$ (7): REM  RING
       BELL
3800   PRINT "DO YOU WANT A NEW FR
       EQ SWEEP ?"
3820   GET Z$
3840   PRINT : IF Z$ = "Y" THEN  GOTO
       3000
3850   PRINT "DO YOU WANT TO CHOP
       OFF BOTTOM OF PLOT? (Y OR N)
       "
3852   INPUT Y$: IF Y$ = "Y" THEN
       PRINT "ENTER CUTOFF LEVEL I
       N NEG. DB": INPUT DZ
3853   IF Y$ = "Y" THEN DR = DH -
       DZ
3860   PRINT "TYPE 'RUN PLOT' (RET
       URN) TO GET A PLOT"
3873   PRINT D$;"OPEN PLOTDATA"
3874   PRINT D$;"WRITE PLOTDATA"
3875   PRINT F2: PRINT DH: PRINT D
       R
3876   FOR I = 1 TO F2
3877   IF DN(I) < DZ THEN DN(I) =
       DZ
3878   PRINT AB(I): PRINT DN(I): PRINT
       PN(I)
3879   NEXT I
3880   PRINT D$;"CLOSE"
3881   END
3900   IF I = 0 THEN 4000
3920 A(I,I) = A(I,I) + V
3940   IF J = 0 THEN 4020
3960 A(I,J) = A(I,J) - V
3980 A(J,I) = A(J,I) - V
4000 A(J,J) = A(J,J) + V
4020   IF I < N THEN 4060
```

```
4040 N = I
4060  IF J < N THEN 4100
4080 N = J
4100  RETURN
4120  IF I = 0 THEN 4220
4140 B(I,I) = B(I,I) + V
4160  IF J = 0 THEN 4020
4180 B(I,J) = B(I,J) - V
4200 B(J,I) = B(J,I) - V
4220 B(J,J) = B(J,J) + V
4240  GOTO 4020
4260  IF I = 0 THEN 4360
4280 B1(I,I) = B1(I,I) + V
4300  IF J = 0 THEN 4020
4320 B1(I,J) = B1(I,J) - V
4340 B1(J,I) = B1(J,I) - V
4360 B1(J,J) = B1(J,J) + V
4380  GOTO 4020
4400  IF I <  > 0 AND K <  > 0 THEN
     A(I,K) = A(I,K) + V
4420  IF J <  > 0 AND L <  > 0 THEN
     A(J,L) = A(J,L) + V
4440  IF J <  > 0 AND K <  > 0 THEN
     A(J,K) = A(J,K) - V
4460  IF I <  > 0 AND L <  > 0 THEN
     A(I,L) = A(I,L) - V
4480  IF K < N THEN 4520
4500 N = K
4520  IF L < N THEN 4560
4540 N = L
4560  GOTO 4020
4580  REM   ### DETERMINANT COMPU
     TATION ###
4600  IF N > 1 THEN 4640
4620 D1 = A(N,N):D2 = B(N,N): RETURN

4640 D1 = 1:D2 = 0:K = 1
4660 L = K
4680 S =  ABS (A(K,K)) +  ABS (B(
     K,K))
4700  FOR I = K TO N
4720 T =  ABS (A(I,K)) +  ABS (B(
     I,K))
4740  IF S >  = T THEN 4780
4760 L = I:S = T
4780  NEXT I
4800  IF L = K THEN 4960
4820  FOR J = 1 TO N
4840 S =  - A(K,J)
4860 A(K,J) = A(L,J)
4880 A(L,J) = S
4900 S1 =  - B(K,J)
4920 B(K,J) = B(L,J):B(L,J) = S1
4940  NEXT J
4960 L = K + 1
4980  FOR I = L TO N
5000 S1 = A(K,K) * A(K,K) + B(K,K
     ) * B(K,K)
5020 S = (A(I,K) * A(K,K) + B(I,K
     ) * B(K,K)) / S1
```

```
5040 B(I,K) = (A(K,K) * B(I,K) -
     A(I,K) * B(K,K)) / S1
5060 A(I,K) = S: NEXT I
5080 J2 = K - 1
5100  IF J2 = 0 THEN 5220
5120  FOR J = L TO N
5140  FOR I = 1 TO J2
5160 A(K,J) = A(K,J) - A(K,I) * A
     (I,J) + B(K,I) * B(I,J)
5180 B(K,J) = B(K,J) - B(K,I) * A
     (I,J) - A(K,I) * B(I,J)
5200  NEXT I: NEXT J
5220 J2 = K:K = K + 1
5240  FOR I = K TO N
5260  FOR J = 1 TO J2
5280 A(I,K) = A(I,K) - A(I,J) * A
     (J,K) + B(I,J) * B(J,K)
5300 B(I,K) = B(I,K) - B(I,J) * A
     (J,K) - A(I,J) * B(J,K)
5320  NEXT J: NEXT I
5340  IF K < > N THEN 4660
5360 L = 1
5380 J2 =  INT (N / 2)
5400  IF N = 2 * J2 THEN 5480
5420 L = 0
5440 D1 = A(N,N)
5460 D2 = B(N,N)
5480  FOR I = 1 TO J2
5500 J = N - I + L
5520 S = A(I,I) * A(J,J) - B(I,I)
     * B(J,J)
5540 S1 = A(I,I) * B(J,J) + A(J,J
     ) * B(I,I)
5560 T = D1 * S - D2 * S1
5580 D2 = D2 * S + D1 * S1
5600 D1 = T
5620  NEXT I
5640  RETURN
5660 N1 = N:N = N - 1:I = 0
5680  FOR K = 1 TO N
5700  IF K < > D1 THEN 5740
5720 I = 1
5740 J = 0
5760  FOR L = 1 TO N
5780  IF L < > D2 THEN 5820
5800 J = 1
5820 A(K,L) = P(K + I,L + J)
5840 B(K,L) = W * Q(K + I,L + J) +
     Q1(K + I,L + J) / W
5860  NEXT L: NEXT K
5880  GOSUB 4600
5900 N = N1
5910  IF ( ABS (D1) +  ABS (D2)) >
     1E18 THEN B1 = 1E20 *  SQR (
     (D1 * 1E - 20) ^ 2 + (D2 * 1
     E - 20) ^ 2): GOTO 5940
5915  IF ( ABS (D1) +  ABS (D2)) <
     1E - 16 THEN B1 = 1E - 20 *
     SQR ((D1 * 1E20) ^ 2 + (D2 *
     1E20) ^ 2): GOTO 5940
```

```
5920 B1 =  SQR (D1 ^ 2 + D2 ^ 2)
5940  IF D1 <  > 0 THEN 6020
5960  IF D2 = 0 THEN 6100
5980  IF D2 > 0 THEN D2 = 90: GOTO
      6100
6000 D2 =  - 90: GOTO 6100
6020  IF D1 < 0 THEN Q = 180: GOTO
      6060
6040 Q = 0
6060  IF D2 < 0 THEN Q =  - Q
6080 D2 = Q + 180 *  ATN (D2 / D1
      ) / PI
6100  RETURN
6200  REM   INFO ABOUT PROGRAM
6220  PRINT "DATA MUST BE IN THE
      FOLLOWING FORMAT:"
6240  PRINT "FOR INSTANCE, FOR AN
      INDUCTOR:"
6245  PRINT "L2 (RETURN)"
6246  PRINT "2,4,.01 (TYPE LOWEST
      NODE FIRST), RETURN"
6260  PRINT
6280  PRINT "R       (RESISTOR)"
6300  PRINT "  FROM NODE #, TO NO
      DE #, VALUE IN OHMS"
6320  PRINT "C       (CAPACITOR)"
6340  PRINT "  FROM NODE #, TO NO
      DE #, VALUE IN UFD"
6360  PRINT "L       (INDUCTOR)"
6380  PRINT "  FROM NODE #, TO NO
      DE #, HENRIES"
6400  PRINT "F       (FET)"
6420  PRINT "  GATE,SOURCE,DRAIN,
      GAIN (A/V)"
6440  PRINT "N       (NPN)
6460  PRINT "  BASE,EMIT,COLL,BET
      A,B-E OHMS"
6480  PRINT "A       (OP AMP)"
6500  PRINT " +IN, -IN, +OUT, -OU
      T, GAIN, OHMS OUT"
6520  PRINT "E        (END OF FILE
      MARKER)"
6540  PRINT "  INPUT NODE #, OUTP
      UT NODE #"
6560  PRINT
6580  RETURN
```

B.2 PLOT

```
10   DATA   1.1,1.2,1.3,1.4,1.5,1.
     7,1.8,2,2.2,2.5,2.7,3,3.5,4,
     4.5,5,5.5,6,6.5,7,7.5,8,9,1
20   DIM AB(200),DN(200),PN(200),X
     X(40),XN(40)
30   F$ = "PLOTDATA"
40   D$ =  CHR$ (4)
50   PRINT D$;"OPEN";F$
60   PRINT D$;"READ";F$
```

```
70   HGR : HCOLOR= 3
80   INPUT F2: INPUT DH: INPUT DR
90 Y = 20
100  IF DR < 200 THEN Y = 10
110  IF DR < 100 THEN Y = 5
120  IF DR < 50 THEN Y = 2
130  IF DR < 25 THEN Y = 1
140  IF DR < 10 THEN Y = .5
150  IF DR < 5 THEN Y = .2
160  IF DR < 2 THEN Y = .1
170  N = Y * ( INT (DH / Y) + 1)
180  N = N - Y
190 NH =   INT ((DH - N) / DR * 15
     9 + .5)
200  IF NH > 159 GOTO 220
210  HPLOT 0,NH TO 279,NH: GOTO 1
     80
220  FOR J = 1 TO F2
230  INPUT AB(J): INPUT DN(J): INPUT
     PN(J)
240  NEXT J
250  PRINT D$;"CLOSE";F$
260 RA = AB(F2) / AB(1):X = 8
270  IF RA < 200 THEN X = 4
280  IF RA < 5 THEN X = 2
290  IF RA < 2 THEN X = 1
300 DD =   INT ( LOG (AB(F2)) /  LOG
     (10))
310 LA =   INT (AB(F2) * 10.^ (3 -
     DD) + .5) * 10 ^ (DD - 3)
320 M = .0001
330 I = 1
340  FOR IX = 1 TO X
350  READ DX: IF DX = 1 THEN   RESTORE
     :M = M * 10
360  NEXT IX
370 DX = DX * M: IF DX < AB(1) THEN
     340
380 XN(I) = DX / M
390 DX =   INT (279 *  LOG (DX / A
     B(1)) /  LOG (AB(F2) / AB(1)
     ) + .5): IF DX > 279 THEN XN
     (I) = 0: GOTO 460
400 XX(I) =   INT (DX / 7 + .5)
410  IF XX(I) < 1 THEN XX(I) = 1
420  IF XX(I) > 38 THEN XN(I) = 0

430 I = I + 1
440  HPLOT DX,0 TO DX,159
450  GOTO 340
460 DN(1) =   INT (159 * ((DH - DN
     (1)) / DR) + .5)
470  HPLOT 0,DN(1)
480  FOR I = 2 TO F2
490 AB(I) =   INT (279 *  LOG (AB(
     I) / AB(1)) /  LOG (AB(F2) /
     AB(1)) + .5)
500 DN(I) =   INT (159 * ((DH - DN
     (I)) / DR) + .45)
```

```
510  IF DN(I) > 159 THEN DN(I) =
     159
520  HPLOT  TO AB(I),DN(I)
530  NEXT I
540  PRINT "DO YOU WANT A PHASE P
     LOT ALSO ?": GET A$: IF A$ =
     "Y" THEN  PRINT "6TH LINE FR
     OM TOP IS 0 DEG; INC=45 DEG"
     : GOSUB 670
550  PRINT "DO YOU WANT A PRINT-O
     UT (Y OR N)?         TURN ON P
     RINTER BEFORE ANSWERING 'Y'.
     ": GET A$
560  FOR I = 1 TO 40
570  IF XN(I) = 0 THEN 600
580  PRINT  TAB( XX(I));XN(I);
590  NEXT I
600  PRINT : PRINT "LAST LINE IS
     ";N + Y;" DB (INC=";Y;"DB &
     45DEG)"
610  PRINT "FREQ. RANGE= ";AB(1);
     " TO ";LA;" HZ";
620  IF A$ < > "Y" THEN 660
630  PRINT : PR# 1
640  PRINT  CHR$ (9);"GM"
650  PR# 0
660  END
670 NP =  INT (159 / DR * (DH + Y
     * (5 -  INT (DH / Y) - PN(1
     ) / 45)) + .5)
680  HPLOT 0,NP
690  FOR I = 2 TO F2
700 NP =  INT (159 / DR * (DH + Y
     * (5 -  INT (DH / Y) - PN(I
     ) / 45)) + .5)
710  IF  ABS (PN(I) - PN(I - 1)) >
     90 THEN  HPLOT AB(I),NP: GOTO
     730
720  HPLOT  TO AB(I),NP
730  NEXT I
740  RETURN
```

B.3 ROOT FINDER

```
10   REM   MOORE ROOT FINDER, TRC,
     7/79
20   REM   TAKEN FROM "CIRCUIT DESI
     GN USING PERSONAL COMPUTERS"
     BY THOMAS CUTHBERT,(WILEY)
100  DIM A(35),B(35),Y(35)
200  REM   INPUT POLY DEG & CMPLX
     COEFS
210  PRINT "N= ";: INPUT N
220  IF N > 35 GOTO 210
230  FOR K = 0 TO N
240  PRINT "EXPONENT= ";K;"      I
     NPUT COEFFICIENT:"
```

```
250   PRINT "   REAL PART= ";: INPUT
      A(K)
260   PRINT "   IMAG PART= ";: INPUT
      B(K)
270   NEXT K
1000  REM    REDUCED POLY RE-ENTRY

1010  IF N = 1 GOTO 5500
1020  YS = 1:X(0) = 1:Y(1) = 1
1030  XS = .1:X(1) = .1
1040  Y(0) = 0:L = 0
1050  GOSUB 3000
2000  REM    NEW X,Y CORNER (LINEA
      R SEARCH)
2010  FS = F
2020  L = L + 1
2030  M = 0:UX = 0:VX = 0
2040  FOR K = 1 TO N
2050  UX = UX + K * (A(K) * X(K -
      1) - B(K) * Y(K - 1))
2060  VX = VX + K * (A(K) * Y(K -
      1) + B(K) * X(K - 1))
2070  NEXT K
2080  PM = UX * UX + VX * VX
2090  DX =  - (U * UX + V * VX) /
      PM
2100  DY = (U * VX - V * UX) / PM
2190  REM    POST QRTRG CUTBACK RE
      -ENTRY
2195  M = M + 1
2200  X(1) = XS + DX
2210  Y(1) = YS + DY
2220  GOSUB 3000
2230  IF F >  = FS GOTO 4000
2240  IF  ABS (DX) > 1E - 5 GOTO
      2260
2250  IF  ABS (DY) <  = 1E - 5 GOTO
      4500
2260  IF L > 50 GOTO 5200
2270  XS = X(1):YS = Y(1)
2280  GOTO 2000
3000  REM    CALC X(.),Y(.),U,V, &
      F
3010  X2 = X(1) * 2
3020  XY = X(1) * X(1) + Y(1) * Y(
      1)
3030  FOR K = 2 TO N
3040  X(K) = X2 * X(K - 1) - XY *
      X(K - 2)
3050  Y(K) = X2 * Y(K - 1) - XY *
      Y(K - 2)
3060  NEXT K
3070  U = 0:V = 0
3080  FOR K = 0 TO N
3090  U = U + A(K) * X(K) - B(K) *
      Y(K)
3100  V = V + A(K) * Y(K) + B(K) *
      X(K)
3110  NEXT K
3120  F = U * U + V * V
```

```
3130   RETURN
4000   REM    FNCN INCRSD SO CUT BA
       CK THE STEP
4010   IF M > 10 GOTO 4040
4020 DX = DX / 2:DY = DY / 2
4030   GOTO 2190
4040   REM    TEST FOR CONVERG > 10
       CUTBACKS
4050   IF  ABS (U) > 1E - 4 GOTO 4
       070
4060   IF  ABS (V) < = 1E - 4 GOTO
       4500
4070   PRINT "STEP SIZE TOO SMALL"

4080   STOP
4500   REM    CONVERGED. PRNT ROOT.
       REMOVE FACTOR.
4510   GOSUB 5000
4520   REM    REMOVE LINEAR FACTOR
4530 K = N - 1
4540 A(K) = A(K) + A(K + 1) * X(1
     ) - B(K + 1) * Y(1)
4550 B(K) = B(K) + A(K + 1) * Y(1
     ) + B(K + 1) * X(1)
4560 K = K - 1
4570   IF K > = 0 GOTO 4540
4580   FOR K = 1 TO N
4590 A(K - 1) = A(K)
4600 B(K - 1) = B(K)
4610   NEXT K
4620 N = N - 1
4630   GOTO 1000
5000   REM    PRNT ROOT
5010   PRINT "A ROOT HAS"
5020   PRINT "    REAL PART= ";X(1)

5030   PRINT "    IMAG PART= ";Y(1)

5040   RETURN
5200   PRINT "NO ROOT FOUND"
5210   STOP
5500   REM    CALC DEG=1 EQUATION
       ROOT
5510   XY = A(1) * A(1) + B(1) * B(
       1)
5520   X(1) = - (B(0) * B(1) + A(0
       ) * A(1)) / XY
5530   Y(1) = (A(0) * B(1) - A(1) *
       B(0)) / XY
5540   GOSUB 5000
5560   END
```

B.4 PARTIAL FRACTION EXPANSION

```
550 AR(J1) = (DR * YR + DI * YI) /
    Y2
560 AI(J1) = (DI * YR - DR * YI) /
    Y2
570   NEXT J
```

```
,
600   NEXT I
610   PRINT "THE  (REAL,IMAG) POLE
      RESIDUES IN"
620   PRINT "INPUT ORDER AND DESCE
      NDING MULTIPLICITY: "
630   FOR I = 1 TO N
635 K = N + 1 - I
640   PRINT "#";I,AR(K),AI(K)
650   NEXT I
660   END


10    REM   PARTIAL FRAC EXPAN, CTI
      /77P44.TRC8/79
20    REM      TAKEN FROM P.433 OF "
      CIRCUIT DESIGN USING PERSONA
      L COMPUTERS", BY THOMAS CUTH
      BERT (WILEY)
100   PRINT "DENOMINATOR DEGREE= "
      ;: INPUT N
110   DIM AR(N),AI(N),PR(N),PI(N)
120   PRINT "INPUT REAL NUMERATOR
      COEFFICIENTS: "
130   FOR I = 1 TO N
140 K = N + 1 - I
150   PRINT "EXPONENT= ";I - 1;: INPUT
      AR(K)
160   AI(K) = 0
170   NEXT I
180   PRINT "INPUT DENOMINATOR ROO
      TS IN ORDER OF "
190   PRINT "ASCENDING MAGNITUDES:
      "
200   FOR I = 1 TO N
210   PRINT "ROOT #";I
220   PRINT "    REAL PART= ";: INPUT
      PR(I)
230   PRINT "    IMAG PART= ";: INPUT
      PI(I)
240   NEXT I
250   EP = 1.E - 10
300   FOR I = 1 TO N
310 I1 = N - I + 1
320   IF I1 = 1 GOTO 400
330   FOR J = 2 TO I1
340 AR(J) = AR(J) + PR(I) * AR(J -
      1) - PI(I) * AI(J - 1)
350 AI(J) = AI(J) + PI(I) * AR(J -
      1) + PR(I) * AI(J - 1)
360   NEXT J
400   FOR J = 1 TO I
410 J1 = N - J + 1
420   IF J = 1 GOTO 460
430   IF (PR(J) - PR(J - 1)) ^ 2 +
      (PI(J) - PI(J - 1)) ^ 2 < =
      EP GOTO 530
440 AR(I1) = AR(I1) - AR(J1 + 1)
450 AI(I1) = AI(I1) - AI(J1 + 1)
```

```
460   IF (PR(J) - PR(I)) ^ 2 + (PI
      (J) - PI(I)) ^ 2 < = EP GOTO
      600
470 YR = PR(J) - PR(I)
480 YI = PI(J) - PI(I)
490 Y2 = YR * YR + YI * YI
500 AS = (AR(J1) * YR + AI(J1) *
      YI) / Y2
510 AI(J1) = (AI(J1) * YR - AR(J1
      ) * YI) / Y2
515 AR(J1) = AS
520   GOTO 570
530 DR = AR(J1) - AR(J1 + 1)
540 DI = AI(J1) - AI(J1 + 1)
```

B.5 Bu/Ch FILTER

```
100   REM  *** BU & CH FILTER DESI
      GN ***
110 PI = 3.1415926: HOME
120   DIM P(80)
130   PRINT "THIS PROGRAM WILL DES
      IGN BUTTERWORTH  ORCHEBYSHEV
      FILTERS, EITHER  LOW-PASS,
        HIGH-PASS, BAND-PASS, OR
      BAND-REJECT.   ANY FREQUENCY
      OR IMPEDANCE LEVEL CAN BE S
      PECIFIED."
140   PRINT : PRINT : INPUT "ENTER
      BUTTERWORTH OR CHEBYSHEV BY
      TYPING 'BU' OR 'CH': ";BC$
150   IF BC$ <  > "BU" AND BC$ <  >
      "CH" THEN   GOTO 140
160   PRINT : INPUT "TYPE 'LP', 'H
      P', 'BP', OR 'BR' TO INDICAT
      E TYPE OF RESPONSE: ";FR$
170   PRINT : INPUT "LOAD RESISTAN
      CE (IN OHMS)= ";R2:R1 = R2
180   PRINT : INPUT "ORDER= ";N: PRINT

190 EO = 1:J =   INT (N / 2): IF N
      / 2 = J THEN EO = 0
200   IF FR$ = "LP" OR FR$ = "HP" THEN
        INPUT "ENTER CUT-OFF FREQUE
      NCY IN HZ: ";F:W = 2 * PI *
      F
210   IF FR$ = "BP" OR FR$ = "BR" THEN
        INPUT "ENTER CENTER FREQUEN
      CY IN HZ: ";F: PRINT : INPUT
      "ENTER BANDWIDTH IN HZ: ";BW
      :W = 2 * PI * F:B = 2 * PI *
      BW
220   PRINT : INPUT "ENTER VARIATI
      ON IN PASS-BAND, IN DB: ";DB
      : PRINT
230 E =   SQR (10 ^ (DB / 10) - 1)
```

```
240   IF BC$ = "BU" THEN   GOSUB 10
      00
250   IF BC$ = "CH" THEN   GOSUB 11
      00
400   REM   ***OUTPUT ROUTINE***
410   PRINT "CIRCUIT VALUES ARE IN
      OHMS, HENRIES, AND FARADS": 
      PRINT
420   PRINT "THIS IS A ";BC$;", ";
      FR$;", ";DB;" DB, FILTER (OR
      DER ";N;")"
440   PRINT : PRINT "THE TOPOLOGY
      AND VALUES ARE SHOWN BELOW.(
      HIT ANY KEY TO CONTINUE THE
      OUTPUT).": PRINT
450   PRINT "<-VIN->": PRINT "¦
       ¦": PRINT "¦      R      R(1
      )= ";R1;" OHMS": PRINT "¦
       ¦"
500   FOR K = 2 TO 2 * J + 2 STEP
      2
510   IF K = N + 2 THEN   GOTO 760
520   IF FR$ = "LP" THEN   PRINT "¦
       ¦": PRINT "¦      L
      L(";K - 1;")= ";P(K - 1) * R
      1 / W;" H": PRINT "¦      ¦"
530   IF FR$ = "HP" THEN   PRINT "¦
       ¦": PRINT "¦      C
      C(";K - 1;")= ";1 / W / P(K -
      1) / R1;" F": PRINT "¦      ¦
      "
540   IF FR$ = "BP" THEN   PRINT "¦
       ¦": PRINT "¦      L
      L(";K - 1;")= ";P(K - 1) * R
      1 / B;" H": PRINT "¦      ¦":
      PRINT "¦      C      C(";K -
      1;")= ";B / W / W / P(K - 1)
      / R1;" F"
545   IF FR$ = "BP" THEN   PRINT "¦
       ¦"
550   IF FR$ = "BR" THEN   PRINT "¦
       ¦": PRINT "¦      ---": PRINT
      "¦    ¦ ¦     C(";K - 1;")= "
      ;1 / B / P(K - 1) / R1;" F"
555   IF FR$ = "BR" THEN   PRINT "¦
      L C": PRINT "¦    ¦ ¦
      L(";K - 1;")= ";B * P(K - 1
      ) * R1 / W / W;" H": PRINT "
      ¦    ---": PRINT "¦      ¦"
610   IF K = N + 1 THEN   GOTO 760
620   IF FR$ = "LP" THEN   PRINT "¦
       ¦": PRINT "*--C--*
      C(";K;")= ";P(K) / W / R1;"
      F": PRINT "¦      ¦"
630   IF FR$ = "HP" THEN   PRINT "*
      --L--*      L(";K;")= ";R1 /
      W / P(K);" H": PRINT "¦
       ¦"
640   IF FR$ = "BP" THEN   PRINT "¦
       ¦": PRINT "*--C--*
```

```
      C(";K;")= ";P(K) / R1 / B;"
      F": PRINT "¦      ¦": PRINT "
      *--L--*      L(";K;") = ";B /
      W / W / P(K) * R1;" H": PRINT
      "¦      ¦"
650   IF FR$ = "BR" THEN  PRINT "¦
         ¦      C(";K;")= ";B * P
      (K) / R1 / W / W;" F": PRINT
      "*-C-L-*": PRINT "¦      ¦
        L(";K;")= ";R1 / B / P(K);
      " H"
750   GET A$
760   NEXT K
800   PRINT "¦      ¦": PRINT "*--R
      --*      R2= ";R2;" OHMS": PRINT
      "<-VOUT>"
810   PRINT
820   IF EO = 0 AND BC$ = "CH" THEN
       PRINT "SINCE THIS IS AN EVE
      N ORDER CHEBYSHEV,  THE TERM
      INATION RESISTANCES CAME OUT
          UNEQUAL. "
900   PRINT : INPUT "DO YOU WANT T
      O DESIGN ANOTHER ONE ?      A
      NSWER BY TYPING 'Y' OR 'N':
      ";A$: IF A$ = "Y" THEN  GOTO
      140
950   END
1000   FOR K = 1 TO N
1010  P(K) = 2 * E ^ (1 / N) *  SIN
      (PI * (2 * K - 1) / 2 / N)
1020   NEXT K
1030   RETURN
1100  H = (1 / E +  SQR (1 + 1 / E
      / E)) ^ (1 / N)
1110  PS = H - 1 / H
1120   IF EO = 0 THEN R1 = R2 / (1
       + 2 * E * E + 2 * E *  SQR
      (1 + E * E))
1130  P(1) = 4 *  SIN (PI / 2 / N)
      / PS
1140   IF EO = 1 THEN J = (N - 1) /
      2
1150   FOR K = 1 TO J
1160  P(2 * K) = 16 *  SIN (PI * (
      4 * K - 3) / 2 / N) *  SIN (
      PI * (4 * K - 1) / 2 / N) /
      (PS * PS + 4 * ( SIN (PI * (
      2 * K - 1) / N)) ^ 2) / P(2 *
      K - 1)
1170   IF K = N / 2 THEN  GOTO 119
      0
1180  P(2 * K + 1) = 16 *  SIN (PI
      * (4 * K - 1) / 2 / N) *  SIN
      (PI * (4 * K + 1) / 2 / N) /
      (PS * PS + 4 * ( SIN (PI * 2
      * K / N)) ^ 2) / P(2 * K)
1190   NEXT K
1200   RETURN
```

answers to odd-numbered problems

1.1. (a) $-2u(t)$ **(b)** $-3u(t+4)$ **(c)** $u(t) - 2u(t-3) + u(t-6)$

1.3. $u(t) - 2u(t-1) + 2u(t-2) - 2u(t-3) + 2u(t-4) - 2u(t-5) + 2u(t-6) - 2u(t-7) + u(t-8)$

1.5.

(a)

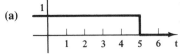

(b)

(c)

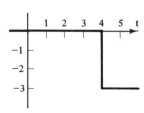

1.7.

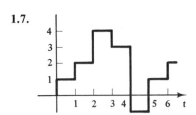

1.9. **(a)** $2 - \dfrac{2t}{3}$ **(b)** $-2tu(-t)$ **(c)** $\left(2 - \dfrac{2t}{3}\right)u(t)$ **(d)** $\left(2 - \dfrac{2t}{3}\right)u(3 - t)$

1.11. **(a)** $2[(1 - t)u(t) + u(t - 1) + u(t - 2) + (t - 3)u(t - 3)]$
    (b) $2t[u(t) - u(t - 1) + u(t - 2) - u(t - 3)]$
    (c) $t[u(t) - 2u(t - 1) + 2u(t - 2) - u(t - 3)]$
    (d) $0.5(t - 2)u(t - 2) - (t - 4)u(t - 4) + (t - 8)u(t - 8) - 0.5(t - 10)u(t - 10)$

1.13.

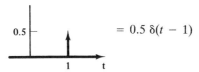

1.15. $(-0.5 \cos 2t)\left[u(t) - u\left(t - \dfrac{9\pi}{4}\right)\right] + 0.5u(t)$

1.17.

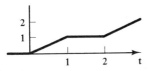 $= 0.5\,\delta(t - 1)$

1.19. **(a)** e^{-2s} **(b)** $4e^{-2s}$ **(c)** $\dfrac{5e^{-3s}}{s}$

1.23. $\dfrac{3535 - 70.7s}{s^2 + 2500}$

1.25. **(a)** $\dfrac{4}{s^3} - \dfrac{1}{s + 1}$ **(b)** $\dfrac{72}{s^5} - \dfrac{12}{s^4} + \dfrac{4}{s + 3} - \dfrac{10}{s^2 + 25} + \dfrac{3s}{s^2 + 4}$

    (c) $\dfrac{24}{s^5} + \dfrac{4}{s^3} + \dfrac{1}{s}$ **(d)** $\dfrac{2}{(s - 1)^3} + \dfrac{4}{(s - 1)^2} + \dfrac{4}{s - 1}$

1.27. **(a)** $\dfrac{6}{(s + 3)^4}$ **(b)** $\dfrac{s + 1}{s^2 + 2s + 5}$ **(c)** $\dfrac{8}{s^2 - 6s + 25}$ **(d)** $\dfrac{20 - 4s}{s^2 - 4s + 20}$

1.29. **(a)** $\dfrac{t^4}{24}$ **(b)** $-4e^{4t/3}$ **(c)** $0.75 \cos (2.5t) - 0.8 \sin (2.5t)$
    (d) $e^{-3t/2}(-0.625t + 0.75)$ **(e)** $(1 - t)4e^{-4t}$

1.31. **(a)** $-\dfrac{1}{6}e^{-t} - \dfrac{4}{3}e^{2t} + \dfrac{7}{2}e^{3t}$ **(b)** $4e^{3t} - e^{-t}$ **(c)** $e^{2t}(6 \cos 4t + 2 \sin 4t)$

1.33. **(a)** $1 - t + \dfrac{t^2}{2} - e^{-t}$ **(b)** $\dfrac{t^3}{9} + \dfrac{t^2}{18} - \dfrac{t}{27} + \dfrac{1}{81} - \dfrac{1}{81}e^{-3t}$

    (c) $1 - e^{-t}\left(1 + t + \dfrac{t^2}{2}\right)$

1.35. **(a)** $2e^t - 2 \cos t + \sin t$ **(b)** $\dfrac{1}{27}(24 + 120t + 30 \cos 3t + 50 \sin 3t)$

    (c) $\dfrac{t}{2}e^{-t} \sin t$

1.37. $\dfrac{2}{s - 1} + \dfrac{-1 + 0.5j}{s + j} + \dfrac{-1 - 0.5j}{s - j} = \dfrac{2}{s - 1} + \dfrac{-2s + 1}{s^2 + 1}$

1.39. $\dfrac{s^2 + 6s + 14}{s^3 + 3s^2}$

1.41. (a) $[2e^{-2(t-2)} - e^{-(t-2)}]u(t - 2)$ (b) $0.5e^{t-3}[\sin(2t - 6)]u(t - 3)$

1.43. $\dfrac{2}{s^2 - 4}$

1.45. $f(0+) = 0$, but $f(\infty)$ cannot be found by this method (see footnote to Eq. 1.26)

1.47. (a) $f(0+) = 0$; $f(\infty)$ doesn't exist   (b) $f(0+) = \infty$; $f(\infty)$ doesn't exist

2.1. (a) $2\dfrac{di}{dt} = 3$; $i(t) = \tfrac{3}{2}t$ (b) $2sI(s) = \dfrac{3}{s}$; $i(t) = \tfrac{3}{2}t$

2.3. (a)

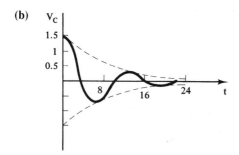

(b) $V_c(s) = \dfrac{1}{s^2 + s/6}$ (c) $v_c(t) = 6 - 6e^{-t/6}$

2.5. (a) $v_c(t) = (6t - 126 + 126.15e^{-t/20})u(t - 1)$

2.7. (a) $v_r(t) = e^{-t/8}\left(\dfrac{3}{2}\cos\dfrac{\sqrt{15}}{8}t - \dfrac{3}{2\sqrt{15}}\sin\dfrac{\sqrt{15}}{8}t\right)$

(b)

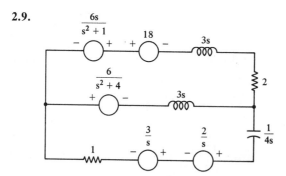

2.9.

2.11. (a) $I_2(s) = \dfrac{-12s + 1}{3s^2 + 3s + 1}$ (b) $i_2(t) = e^{-t/2}\left(-4\cos\dfrac{t}{\sqrt{12}} + \dfrac{7\sqrt{12}}{3}\sin\dfrac{t}{\sqrt{12}}\right)$

2.13. $V_c(t) = 2.48\sin(t + 7.1°)$

2.15. (a) $G(s) = -\dfrac{s^2 + 1}{s^2}$ **(b)** $v_o(t) = -5 - \dfrac{5}{2}t^2$

2.17 $V_o/V_1 = 1/(s + 2)$

2.19. $Z_{dp}(s) = \dfrac{2s^4 + 5s^3 + 7s^2 + 5s + 2}{s^4 + 3s^3 + 4s^2 + 3s + 2}$

2.21.

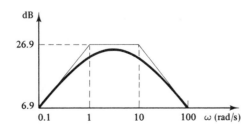

2.23. $y(t) = 10 - 10e^{-t}$

2.25. (a) $-\cos t + 3 \sin t + e^{-3t}$ **(b)**

(b)

(c) $-\dfrac{10}{3}e^{-3t} + \dfrac{10}{3}$

2.27. $G(s) = \dfrac{2s^2 + 15s}{s^2 + 25}$

2.29. $4 + 8t$

2.31. $g(t) = k_1 e^{-t} + k_2 e^{-2t} + k_3 t e^{-2t} + k_4 e^{-3t} \sin (t + \theta)$

2.33. $G(s) = 3\left[\dfrac{(s^2 + 1)(s + 3)}{(s + 2)^2(s + 1)(s^2 + 2s + 2)}\right]$

2.35. $\tau = 0.10$ s; rise time $= 0.22$ s; overshoot $= 0\%$; settling time $= 0.39$ s

2.37. (a) underdamped   **(b)** 211 Hz   **(c)** underdamped   **(d)** 3.54 Ω

2.39. (a) 170 Hz   **(b)** 0.9, which is -0.9 dB

2.41. $G(s) = \dfrac{220s}{s^2 + 11s + 10} = \dfrac{220s}{(s + 10)(s + 1)}$

3.1. (a) BR, passive, analog   **(b)** AP, passive, analog   **(c)** integrator, active, analog   **(d)** BP, active, analog, ceramic

3.3.

f	.25	2.5	25	250
v_o	1.00	.99	.56	.05

This is a low-pass filter.

3.5. **(a)** $\dfrac{1}{s^2 + 2s + 2}$ **(b)** $\dfrac{1}{\sqrt{4 + \omega^4}}$ **(c)** 1.414   **(d)** -9 dB   **(e)** yes, because

$$A(\omega) = \text{Butterworth form} = \frac{k}{\sqrt{1 + (\omega/\omega_0)^{2n}}} \quad (k = 0.5,\ \omega_0 = \sqrt{2};\ n = 2)$$

3.7.

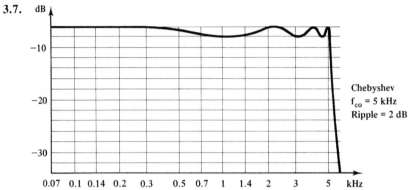

3.9.

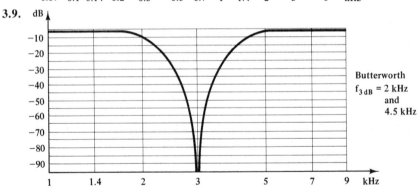

3.11.

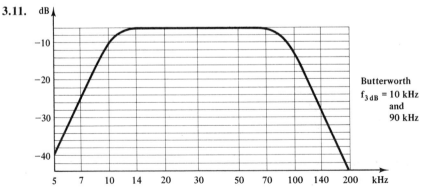

3.13. none

3.15. inverse Chebyshev

4.1.

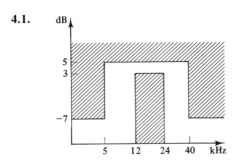

4.3. $n = 4$

4.5. 318 dB/decade

4.7.

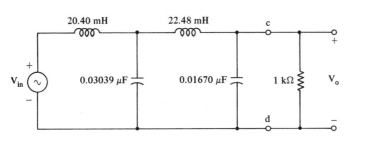

Actual dc gain = 0 dB; to get −8 dB, tap R_L
(replace 1 kΩ at points cd by circuit shown →)

4.9.

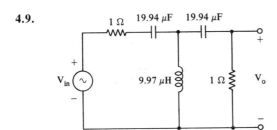

4.11. (a) HP **(b)** Chebyshev **(c)** 0.5 dB, third order **(d)** 2 kHz **(e)** −26 dB

4.13. 30 db/decade or 9 dB/octave

4.15. (a)

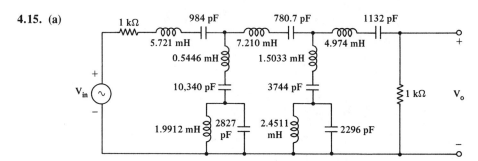

(b)

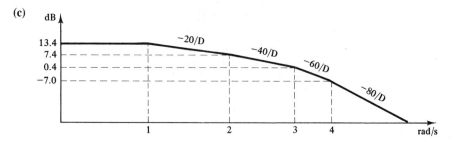

C and L are the same as above

4.17.

5.1. $-\dfrac{1}{4s^2}$

5.3. (a) $Z_T = -\dfrac{1}{s}$ **(b)** negative capacitor $= 1.0$ F

5.5. yes; Y_2 and Y_5 are resistors; Y_1, Y_3, and Y_4 are capacitors; $R_2 = 0.25\ \Omega$, $R_5 = 1\ \Omega$

5.7. (a) $G(s) = \dfrac{8}{s^2 + 3s + 2}\,\dfrac{14}{s^2 + 7s + 12}$

     (b) $G(s) = \dfrac{112}{(s + 1)(s + 2)(s + 3)(s + 4)}$

(c)

dB

13.4 ─────────
7.4
0.4
−7.0

−20/D

−40/D

−60/D

−80/D

1 2 3 4 rad/s

5.9. (a)   **(b)** Chebyshev   **(c)** LP, second order   **(d)** 1 dB   **(e)** 6 kHz

6.1. $G(s) = \dfrac{0.3535 \times 10^6}{s^2 + 1500s + 0.5 \times 10^6} \dfrac{0.707 \times 10^4}{s^2 + 205s + 10^4}$

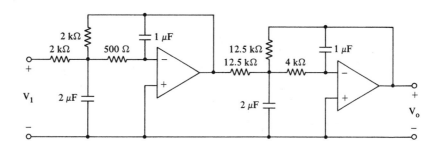

6.3. $G(s) = \dfrac{1.179 \times 10^7}{s^2 + 6094s + 1.179 \times 10^7} \times \dfrac{1.500 \times 10^7}{s^2 + 4226s + 1.5 \times 10^7} \times \dfrac{3315}{s + 3315}$

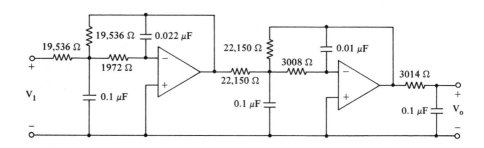

6.5. $G(s) = \dfrac{1.778}{s^2 + 1.8477s + 1} \times \dfrac{1.778}{s^2 + 0.7654s + 1}$ (prototype)

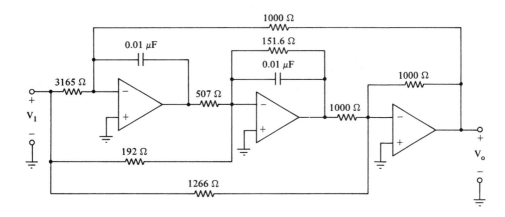

6.7.

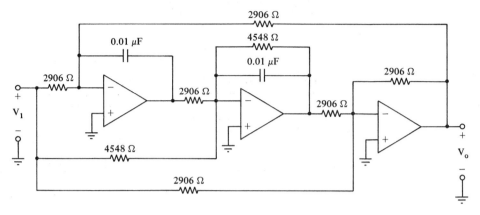

6.9. See page 307.

7.1. **(a)** $i(t) = C_i f_0[2v_1(t) - v_c(t)] \dfrac{C_2}{C_1 + C_2}$ **(b)** 6V

(c)

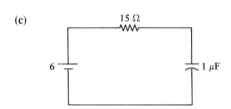

(d) $\dfrac{V_o(s)}{V_1(s)} = \dfrac{1.33 \times 10^5}{s + 0.67 \times 10^5}$

7.3. **(a)** $G(s) = \dfrac{1.974 \times 10^6}{s^2 + 3770s + 1.974 \times 10^6}$

(b)

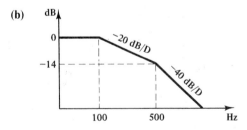

7.5. See page 308.

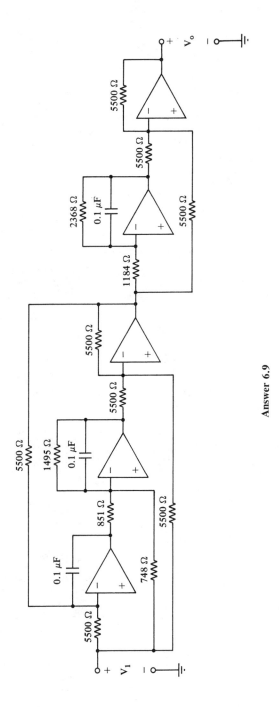

Answer 6.9

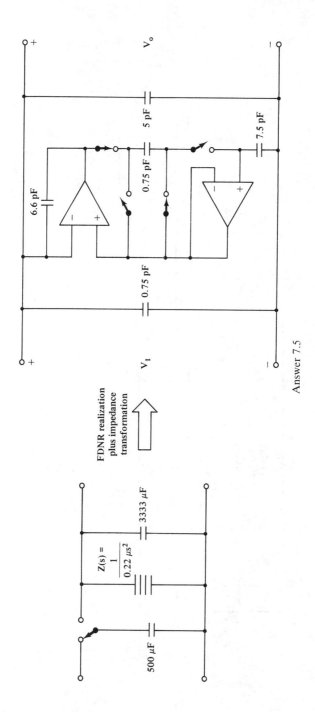

FDNR realization
plus impedance
transformation

$$Z(s) = \frac{1}{0.22 \ \mu s^2}$$

3333 μF

500 μF

6.6 pF

0.75 pF

5 pF

0.75 pF

7.5 pF

V_1

V_o

Answer 7.5

308

7.7. (a) $G_{LP}(s) = \dfrac{1.5162}{s^2 + 1.4256s + 1.5162}$

(b) $G_{BP}(s)(\text{norm.}) = \dfrac{3.4114s^2}{s^4 + 2.1384s^3 + 5.4114s^2 + 2.1384s + 1}$

(c)

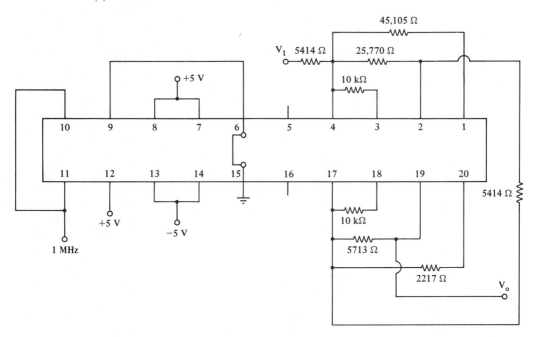

8.1. $\dfrac{3z}{z - 6}$ or $\dfrac{3}{1 - 6z^{-1}}$

8.3. (a) $\left(\dfrac{z + 1}{z}\right)^2$ **(b)** $\left(\dfrac{z + 1}{z}\right)^7$

8.5. $\dfrac{5z^3 + z^2 - 2}{z^3}$

8.7. $\dfrac{z^4 + z^3 - 2z^2 - 2z - 2}{z^5}$

8.9. $\dfrac{z + 1}{z}$

8.11. (a) $\dfrac{z}{(z - 0.5)(3z - 2)}$ **(b)** $-2(0.5)^n + 2(\tfrac{2}{3})^n$ **(c)** 0, .333, .389, .342 **(d)** 0

8.13. (a) unstable **(b)** unstable **(c)** unstable

8.15. (a) $\dfrac{1}{1 - e^{-(s+1)T}}$ **(b)** $\dfrac{z}{z - e^{-T}}$

8.17. (a) $\dfrac{0.951z - 1}{z - 0.951}$ **(b)**

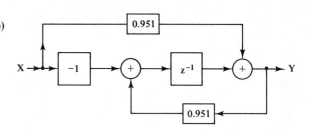

glossary

Active filter: A filter involving amplifiers. It has many advantages, allowing realization of transfer functions not possible with passive circuits.

A/D converter: A device that takes an analog signal as an input and produces a digital signal as its output. This output is proportional to the samples of the analog input.

Adder: An active circuit whose output is a linear combination of two or more input signals.

All-pass filter: A filter whose gain magnitude does not vary with frequency, but whose output phase relationship to the input does vary with frequency.

All-purpose bi-quad: A bi-quad filter whose transfer function can be a quadratic in s divided by another quadratic in s, so that either LP, HP, BP, BR, or AP filters can be realized; but to switch from one to the other, components must be changed.

Analog sampled data: Data that consist of periodic samples of an analog signal. Each sample is then held during its period. This train of samples is then processed in an analog manner. *See also* Switched capacitor.

Asymptote: A straight line on a Bode plot which more or less approximates the true response over a certain range of frequencies. It is found by making certain simplifying assumptions in the frequency range of interest.

Attenuation: The amount of loss of signal, expressed in decibels, between the input and output of a circuit. It is the negative of the dB gain value.

Band-elimination filter: *See* Band-reject filter.

Band-pass filter: A filter whose output remains relatively high if the input is between two given frequencies, but is low when the frequency is outside the band defined by these two frequencies.

Band-reject filter: A filter whose output remains relatively low if the input is between two given frequencies, but is high when the frequency is outside the band defined by these two frequencies.

Band-stop filter: *See* Band-reject filter.

Bandwidth: The width, in radians/second or in hertz, of a region of the frequency response curve in which the output is either above or below a given percentage, usually (but not always) 70.7% of the peak output.

Bessel filter: A low-pass monotonic filter that for a given order has a gain magnitude plot that is inferior to a Butterworth filter. However, it has a superior phase response. Its phase changes linearly with frequency over most of the passband.

Bi-quad filter: An active state-variable filter that has a transfer function whose denominator is a quadratic in s, and whose individual active sections bear a direct relation to the operations corresponding to the powers of s in the transfer function.

Blend point: *See* Breakpoint.

Breakpoint: The frequency at which two consecutive asymptotes cross each other.

Brick-wall filter: An ideal filter that can never be realized, but whose response is 100% in the passband and is 0% everywhere else.

Bruton transformation: Dividing or multiplying all element impedances of a circuit by s. This will not change the frequency response, but it could, for example, change all inductors to resistors, all resistors to capacitors, and so on.

Buffer: An active device whose transfer function is 1 and whose only purpose is to prevent loading of the circuitry preceding the buffer.

Butterworth filter: A type of filter whose response has no ripples and whose gain function is the simplest of all gain functions to write. It is a maximally flat filter.

Cauer filter: *See* Elliptic filter.

Chebyshev filter: A filter containing ripples in the passband. These ripples are of uniform height. This filter drops more quickly than the Butterworth after leaving the passband.

Chebyshev polynomials: A group of polynomials in ω that are used to construct the formula for the gain function, $A(\omega)$, of a Chebyshev filter.

Continuing fraction: A fraction whose numerator or denominator contains another fraction whose numerator or denominator contains another fraction, and so on.

Critically damped: A system which when excited by a step function will have its output rise quickly and not overshoot its final value, although it would overshoot if any of its components were changed by the slightest amount.

Current transfer ratio: The ratio of the output current of a circuit to the input current. It is usually expressed in the s domain.

Cutoff: The frequency at which the output drops below a certain percentage (usually, but not always, 70.7%) of the maximum output value.

D/A converter: A device that takes a digital signal (i.e., a series of successive discrete values) and converts them into a smoothly changing analog signal at the output.

Damped natural frequency: The frequency of the sinusoidal portion of the mathematical expression for the time response of a circuit which is excited by an impulse or step function. This applies to underdamped circuits only.

Damping: Refers to a factor in a time function that is responsible for that function to gradually die out as time goes by.

Damping constant: This constant, determined from the circuit element values, is the reciprocal of the time constant of the decay of oscillation of the underdamped circuit.

Damping ratio: The ratio of damping constant to the undamped natural frequency. If less than 1, the circuit will be underdamped.

Damping theorem: A Laplace theorem that given the Laplace transform of a time function, makes it possible to find the Laplace transform of $e^{-\alpha t}$ times that time function, without ever knowing or having to find either time function.

Darlington realization: A passive filter circuit consisting of a ladder arrangement of capacitors and inductors with a resistor at the load end and possibly a resistor at the source end also.

dB: The decibel is defined as $20 \log_{10} N$, where N is the magnitude of a voltage or a current or any ratio involving voltages or currents that are used in the definition of transfer function. If a power ratio is used, dB is defined as 10 $\log_{10} (P_0/P_1)$.

Decade: A span of frequencies covering a 10:1 frequency ratio.

Delay element: A device or set of computer statements that causes a signal or value to be delayed in the circuit or program.

Differential theorem: A Laplace theorem that given the Laplace transform of a function of time, makes it possible to find the Laplace transform of the derivative of the time function, without knowing the time function or its derivative.

Differentiator: A circuit whose output signal is the derivative of its input signal.

Digital data: A series of discrete values which are usually samples of some analog signal arranged in the time order in which they occur. The values are rounded to the nearest digital bit.

Digital filter: A device that uses digital data as an input and gives digital data as an output with a magnitude and phase change that is frequency dependent.

Digital sample: *See* Digital data.

Digital simulation: A combination of A/D converter, digital filter, and D/A converter that will simulate or be indistinguishable from an analog filter.

Discrete data: A series of samples of a signal, arranged in the order of time in which they occur.

Discriminant: That portion of the quadratic formula that tells whether the root is real or imaginary (i.e., $b^2 - 4ac$).

Double-sided Laplace transform: An alternative type of Laplace transform, not used in this book. The limits of integration in the defining integral are from negative infinity to positive infinity.

Driving-point admittance: The admittance, usually expressed in the *s*-domain, that is connected to a circuit's source.

Driving-point impedance: The impedance, usually expressed in the *s*-domain, that is connected to a circuit's source.

Electromechanical: Description of a device whose operation depends on voltage and current as well as force and speed. Example: electric motor.

Elliptic filter: A filter whose Bode magnitude plot contains equal height ripples in the passband and also the same number of equal height ripples in the stopband.

Equiripple filter: A filter in which the maxima of all the ripples are the same and the minima of all the ripples are the same.

Expand: To multiply all factors in an expression so as to create one factor which is a polynomial.

Expansion as a quadratic: Incomplete partial fraction expansion, in which the denominator of the fraction in question is a quadratic in s. The numerator, when solved, is of the form $As + B$. This method is only used in the case where the roots of the quadratic are complex or imaginary.

Expansion by poles: Partial fraction expansion involving fractions whose denominator is of the form $s + a$. When solved, the numerators will be constants.

Feedback: A feature of an active circuit in which the output signal is used as one of the inputs to the circuit.

FDNR: A frequency-dependent negative resistor is an active circuit that behaves as a resistor except that the current flows the "wrong" way. Also, the value of the resistor changes with frequency.

Filter: A device with an input and an output in which the output magnitude and/or phase depends on the frequency of the input, in addition to depending on the input's magnitude and phase.

Final value theorem: A simple means of finding the value of a time function at t = infinity if the time function is not given but its Laplace transform is.

FIR filter: In a finite impulse response digital filter, if the input is made zero for all values of time after a particular time, the output will become zero some finite time later and stay zero.

Forced response: The output of a circuit if the input is a time function that is changing: for example, a square wave or a sine wave.

Frequency plot: A plot of the frequency response versus frequency.

Frequency response: The variation of the magnitude and phase of the sinusoidal output of a circuit as a function of frequency when the input is a sine wave of zero phase and unity magnitude.

Frequency scaling: A method of finding a circuit's new element values if it is desired to alter a given circuit so that its frequency response will be shifted up or down in frequency.

Frequency shifting theorem: If a circuit or transfer function is to be frequency scaled, this theorem will find the new transfer function if the original transfer function is known.

Gain: The ratio of the magnitude of the steady-state sinusoidal output to the input. This gain is usually converted to dB (*see* dB), so that a positive dB gain means that the output is greater than the input and a negative gain means that the output is less than the input.

Geometrical midpoint: The geometrical mean of two values is $\sqrt{N_1 N_2}$, where N_1 and N_2 are the two values.

GIC: A generalized immittance converter is a device attached to any impedance, causing it to behave as a different impedance. The new impedance will be the old impedance multiplied by the function of s that is the characteristic conversion function of that GIC.

Ground plane: A grounded layer underlying a circuit, such as an integrated circuit. It can serve as a shield and as a ground for those elements in the IC that are grounded.

Group delay: The incremental rate of change of phase (in radians) for the frequency response for a small change in frequency (in rad/s). The unit for group delay is seconds.

Gyrator: A device that will convert any impedance connected to it into one having an impedance equal to the reciprocal of that impedance.

Half-ripples: A portion of a frequency plot between a local maximum and the next local minimum, or vice versa. The number of half-ripples relates to the order of Chebyshev filters.

High-pass filter: A filter whose output is relatively high when the input frequency is above a particular frequency and is low below that frequency.

IIR filter: An infinite impulse response filter is a digital filter in which when the input ceases (goes to zero), the output either never reaches zero or it takes an infinitely long time to settle to zero. It can only be described by a recursive equation.

Impedance conversion constant: The value D of the FDNR. The impedance of the FDNR is $1/(s^2 D)$. Thus, for an FDNR, D plays a role similar to that of C and L in finding the impedance of a capacitor or an inductor.

Impedance scaling: A technique of altering a circuit's element values, so they are common commercially available values or are compatible with the impedance level of other connected circuitry, without changing the frequency response of the circuit.

Impulse: A time function that is everywhere zero, except at one particular point, where it then rises extremely fast to a very high value, then immediately falls back to zero, yet the area under this curve is finite.

Impulse response: The output time function when the input is a unit impulse.

Initial conditions: The values of the currents through the inductors and the values of the voltages across the capacitors in the given circuit at time = zero.

Initial value theorem: A simple means of finding the value of a time function at *t* = zero if the time function is not given but its Laplace transform is.

Input admittance: The ratio of input current to input voltage.

Input impedance: The ratio of input voltage to input current.

Integral theorem: A Laplace theorem that given the Laplace transform of a function of time makes it possible to find the Laplace transform of the integral of the time function, without knowing the time function or its integral.

Integrator: A circuit whose transfer function is $1/s$. A circuit whose output is the integral of its input. This circuit usually takes the form of an operational amplifier with a capacitor in its feedback loop.

Integrodifferential equation: An equation involving both the derivative and the integral of the voltage or current in question.

Inverse Chebyshev filter: A filter containing ripples in the stopband. These ripples are of uniform height. This filter drops more quickly than the Butterworth after leaving the passband.

Inverse Laplace transform: The time function corresponding to a given *s*-domain function, or the process for converting an *s*-domain function into the time function.

Laplace transform: The *s*-domain function corresponding to a given time function, or the process of converting the time function into an *s*-domain function.

Left-half plane: That portion of the complex plane in which the real part of the complex number is negative.

Linearity property: This states that the Laplace transform of any linear combination of time functions is equal to the same linear combination of Laplace transforms of the individual time functions.

Linear section: In an active circuit, that portion whose transfer function is $b/(s + a)$, thus containing only one reactive component.

Lissajous: A pattern formed on an oscilloscope when one circuit voltage controls the *x* direction and another one the *y* direction. Some typical Lissajous patterns happen to be Chebyshev polynomials.

Loading of a filter: The attachment of a load on the filter, so that an undesired change in the frequency response of the filter will occur.

Low-pass filter: A filter whose output is relatively high when the input frequency is below a particular frequency and is low above that frequency.

Marginally stable: The property of a circuit or system which makes it stable for most input frequencies but unstable at a limited number of frequencies.

Maximally flat filter: Of all the types of filters of the same order, this one has the most derivatives of the gain function equal to zero at a particular frequency (at dc for a low-pass filter). *See also* Butterworth filter.

MF-10 IC chip: An integrated circuit featuring switched capacitors. It can be used to build a variety of switched capacitor filters. Higher-order filters can be built by cascading these chips. An external clock and some resistors are also required.

Monotonic: A function which, when plotted, has a slope which is positive over the whole plot, or negative over the whole plot.

MOS switch: A metal-oxide-semiconductor field-effect transistor that is used to simulate a switch. It is operated by applying a suitable dc level to its gate terminal. It is used in switched capacitor circuits.

Multiple complex poles: Poles that are complex numbers, but are repeated, occurring two or more times.

Multiple-feedback filter: An operational amplifier circuit containing elements connected from the output to two or more different points before the op-amp. It is used to realize a second-order filter section.

Multiple real poles: Poles that are real numbers, but are repeated, occurring two or more times.

Multiplicity: The number of times that a repeated root or factor occurs in a given expression.

Natural frequency: The frequency, or frequencies, that occur in the natural response.

Natural response: The time output of a circuit when the input is a natural input, that is, when the input is a unit impulse or a unit step.

Negative capacitor: A device that has an impedance that decreases in magnitude with frequency as a normal capacitor would, except that the current lags the voltage by 90°.

Negative inductor: A device that has an impedance that increases in magnitude with frequency as a normal inductor would, except that the current leads the voltage by 90°.

Negative resistor: A device that behaves like a resistor, except that the current is 180° out of phase with the voltage.

NIC: A negative immittance converter. A device that will convert the impedance of any element connected to it to its negative. *See also* negative capacitor, negative inductor, and negative resistor.

Nomograph: A chart by means of which it is possible to solve graphically particular equations by constructing straight lines between given numbers on the axes of the chart, according to a certain procedure. Nomographs are used in finding the order of a filter that is to meet certain specifications.

Nonrecursive equation: As applied to digital circuits, an equation describing the output in terms of past and present samples of the input only.

Octave: A span of frequencies covering a 2:1 frequency ratio.

Omega: The uppercase Greek letter, Ω, is the frequency ratio used when dealing with nomographs for finding the order of filters. Its definition depends on whether LP, HP, BP, or BR filters are being designed. It is the horizontal axis of these nomographs.

Order: The degree of complexity of a filter. It is equal to the highest power of s in the transfer function.

Output admittance: The reciprocal of output impedance.

Output impedance: The Thévenin impedance, or impedance looking back into the circuit, from the output end.

Overshoot: The amount by which the peak value of the time response to a step input exceeds the final value of the response, expressed as a percent of this final value.

Parallel sections: A method of connecting op-amp sections in which the same input signal is applied to both sections. The two outputs are then added in an op-amp adder. This technique of realizing an active filter is little used.

Parasitic capacitance: An unwanted capacitance in an integrated circuit, caused by the proximity of various layers and conductive paths to each other.

Partial fraction expansion: The technique of writing a function of s, whose numerator and denominator are polynomials in s, as a sum of several simple fractions, each of which can have its inverse Laplace transform easily calculated.

Passband: The band of frequencies in which the gain magnitude of the filter response is above some predetermined amount.

Passive: Describes a component or system that contains no amplifiers.

Phase response: *See* Frequency response.

Phase-time delay: The negative of the phase of the filter output signal (in rad/s) divided by the frequency (in rad/s).

Pole: A root of the denominator polynomial of the s-domain function.

Prototype filter: A low-pass filter that cuts off at 1 rad/s and contains components that have numeric values near 1. This filter is the starting point from which other filters (such as HP, BP, and BR) at other impedance levels and with different cutoff frequencies are developed by various rules.

Pulse: A time function that has a constant value during a given time interval and is zero outside this interval.

Quadratic section: In an active filter, an op-amp section that has a transfer function whose denominator is a second-order polynomial in s.

Quartic: A polynomial whose highest power of the variable is the fourth power.

Quartz: A glasslike mineral that is used to make accurate crystals used in oscillators found in radios and watches.

Ramp: A time function that is zero when off and plots as a slant line when on.

RC–CR transformation: A method of converting a low-pass filter to a high-pass filter, or vice versa. It consists of changing all resistors to capacitors (the R value will be $1/C$) and all capacitors to resistors ($C = 1/R$).

RCRC filter: A higher-order passive ladder-type filter consisting of just resistors and capacitors. Its characteristics are inferior to the more common type of filters, such as Butterworth.

Realization: The process of designing a filter to meet certain specifications or to have a given transfer function.

Recursion formula: A procedure for generating formulas of higher order when the corresponding lower-order formulae are available.

Recursive equation: As applied to digital circuits, an equation that describes the output in terms of past samples of the output as well as past and present samples of the input.

Residue: The numerator constant that is calculated for each fraction in the partial fraction expansion process.

Ripple: The up-and-down motion of the gain magnitude plot of a Chebyshev or elliptic filter response.

Rise time: Regarding the time plot of the step response of a system, rise time is the elapsed time between the point at which the output is at 10% of the final output value and the point at which the output is 90% of final value.

Roll-off rate: The slope of the gain magnitude frequency response plot, when dB is plotted against frequency on semilog graph paper. The proper unit is either dB/decade or dB/octave.

Sallen-Key filter: An active filter that uses a noninverting op-amp and has a special feedback arrangement so that second-order LP and HP and some BP filters can be realized.

Sampling period: The time interval between successive samples in a discrete or digital system.

***s-domain*:** Describing an expression, equation, or quantity in a circuit or system, by using its Laplace transform, so that the letter *s* appears in it, instead of time, *t*.

***Settling time*:** The amount of time it takes for the output to settle to within 2% of its final value from the instant a step is applied at the input of the circuit or system.

***Shifted function*:** A function that is identical in every respect to another function, except that all the values occur later, delayed by the same amount of time.

***Shifting theorem*:** A Laplace theorem that given the Laplace transform of a time function makes it possible to find the Laplace transform of the shifted function, without ever knowing either time function.

***Simple complex poles*:** Poles that are complex numbers and that are not repeated, but do come in complex-conjugate pairs.

***Simple real poles*:** Poles that are real numbers and that are not repeated.

***Single-sided Laplace transform*:** The most common and most useful of the two types of Laplace transform and the only one of concern in this book. The limits of integration are from zero to positive infinity.

***Stability*:** The ability of a circuit's or system's output to remain within bounds, when excited by various bounded inputs.

***Stable*:** The property of a circuit or system that will keep its output within bounds when excited by any bounded input.

***Steady-state response*:** The portion of the forced response or output of a circuit or system when the input pattern repeats periodically (e.g., a sinusoid) and sufficient time has passed so that the output also repeats periodically (e.g., a sine wave)

***Step function*:** A time function whose plot looks like a step; that is, it suddenly jumps from one dc level to another.

***Step response*:** The output time function when the input is a unit step function.

***Stopband*:** The band of frequencies in which the gain magnitude of the filter response is below some predetermined amount.

***Summing amplifier*:** *See* Adder.

***Switched capacitor*:** A capacitor that is periodically switched back and forth between two different locations in the circuit many times per second. The capacitor has approximately the same effect in the circuit as if a resistor were connected permanently between the two locations.

***Synthesis*:** *See* Realization.

Time-scaling theorem: A Laplace theorem that given the Laplace transform of a time function makes it possible to find the Laplace transform of a time-stretched (or slow-motion) version of the original time function without ever knowing either time function.

Topology: The layout of the branches of a circuit without the component values given.

Transconductance: The transfer function of a VCT. The unit is siemens.

Transducer: A device that converts an input into an output, usually from one technology into another; for instance, an electrical signal is converted to a mechanical movement of a meter needle.

Transfer admittance: The ratio of the current at the output of a circuit to the voltage at the source. It is usually expressed in the *s*-domain.

Transfer function: The ratio, in the *s*-domain, of the output of a circuit or system to the input.

Transfer impedance: The ratio of the voltage at the output of a circuit to the current at the source. It is usually expressed in the *s*-domain.

Transform: A change of an expression or equation from one system of calculation to another. Thus logarithms transform multiplication and division to addition and subtraction. Laplace transforms change differential equations into algebraic equations and *z*-transforms change difference equations into algebraic equations.

Transform analysis: The application of transforms to the solutions of circuit or system problems, whereby the input can be a wide variety of time functions.

Transforming of elements: The process of changing the normal component values into quantities that permit their use when writing the *s*-domain equivalent of such formulas as Ohm's law, Kirchhoff's voltage law, and so on.

Transforming of equations: The process of changing differential equations into algebraic *s*-domain equations so as to facilitate their solution.

Transient response: That portion of the circuit output that occurs from the instant that the input is applied until the output has settled into a repetitive or constant steady-state condition.

Transition band: The band of frequencies in which the gain magnitude of a filter response is less than that required to qualify for the passband but more than that required to qualify for the stopband.

Trick formula: A means of taking the inverse Laplace transform of a fraction in the partial fraction expansion of an *s*-domain expression if the pole of that fraction is simple complex without knowing or having to find that fraction's numerator.

***Tuning of filters*:** A procedure for changing the frequency response of a physical circuit slightly by adjusting one or more variable resistors to counteract the effect of a frequency response change due to inaccurate components.

***Undamped natural frequency*:** The frequency of the output of a circuit that is excited by a step or impulse if all the resistors are removed. In an *RLC* circuit excited by a sinusoid, the frequency of the input that will cause the output steady-state magnitude to be a maximum.

***Unit impulse*:** A unit pulse that is very high and very narrow.

***Unit pulse*:** A time function that starts at zero, rises a certain amount, and then drops back to zero at some later time and the area under this plot during this time interval is 1 unit.

***Unit step*:** A step time function that changes its height by 1 unit: for example, 1 volt or 1 ampere.

***Universal filter*:** An active filter that can give a LP, HP, and BP output all at the same time just by designating the proper node of the circuit as the output node.

***Unstable*:** The property of a circuit or system which cannot keep its output within bounds when excited by any input.

***VCT*:** A voltage-to-current transducer, an active device whose output current is proportional to its input voltage, and this output current does not change when the size of the load element connected to the output terminals is changed.

***VCVS*:** A voltage-controlled voltage source, an active device: for instance, a non-inverting op-amp. *See also* Sallen-Key filter.

***Voltage transfer ratio*:** The ratio of the output voltage of a circuit to the input voltage. It is usually expressed in the *s*-domain.

***Zero*:** A root of the numerator polynomial of the *s*-domain function.

***z-transform*:** The *z*-domain function corresponding to a given series of signal samples taken at periodic time intervals, or the process of converting from the sequence of signal samples to the *z*-function. It is a transform process that is very useful when working with discrete or digital data.

bibliography

T. R. CUTHBERT, *Circuit Design Using Personal Computers*, Wiley, New York, 1983.

L. P. HUELSMAN and P. E. ALLEN, *Introduction to the Theory and Design of Active Filters*, McGraw-Hill, New York, 1980.

D. E. JOHNSON, *Introduction to Filter Theory*, Prentice-Hall, Englewood Cliffs, N.J., 1976.

S. KARNI, *Network Theory: Analysis and Synthesis*, Allyn and Bacon, Boston, 1966.

A. PAPOULIS, *Circuits and Systems, a Modern Approach*, Holt, Rinehart and Winston, New York, 1980.

R. STEINCROSS, "BASIC Program Performs Circuit Analysis," *EDN*, pp. 260–261, September 1, 1982.

M. E. VANVALKENBURG, *Analog Filter Design*, Holt, Rinehart and Winston, New York, 1983.

G. H. WARREN, "Computer-Aided Design Program Supplies Low-Pass-Filter Data," *EDN*, pp. 143–150, August 20, 1980.

index